ELECTRICIAN'S LICENSING EXAM

ELECTRICIAN'S LICENSING EXAM

Published in the United States by LearningExpress, LLC, New York.

Library of Congress Control Number: 2009926363

A copy of this title is on file with the Library of Congress.

ISBN: 978-1-57685-684-0

Printed in the United States of America

9 8 7 6 5 4 3 2

First Edition

Regarding the Information in This Book
We attempt to verify the information presented in our books prior to publication. It is always a good idea, however, to double check such important information as the test format, application and testing procedures, and deadlines, as such information may change from time to time.

For more information or to place an order, contact LearningExpress at:
2 Rector Street
26th Floor
New York, NY 10006

Or visit us at:
www.learnatest.com

Acknowledgments
Statements and tables appearing in this book from the *National Electrical Code®* are reprinted or adapted with permission from NFPA 70® and the *National Electrical Code®*. Copyright © 2007, National Fire Protection Association, Quincy, MA 02169. This reprinted material is not the complete and official position of the NFPA on the referenced subject, which is represented only by the standard in its entirety.

About the Authors

Landon Johnson has over twenty years of teaching experience in Career Education as well as nearly thirty years in the field of electronics. Jump-starting his career in the late '70s through Cooperative Education with a variety of opportunities, he received his BSEE in 1985 and began work with a NASA subcontractor in Columbia, Maryland. In 1987, he began a teaching career, which has remained a constant in his professional life in spite of frequent returns to the industry. In 1999, he joined his current employer, YTI Career Institute, as a full-time instructor in the area of Electronics Engineering Technology. He is currently Program Director of Trades and Technology for YTI Career Institute. He resides near York, Pennsylvania, with his wife, Stacy, and their daughter, Savannah.

Allen Stonebraker graduated from Northeastern School District of York, Pennsylvania, in 1969. In 1975, he began doing work as an electrician in York and in 1990 he received his Master Electrician license in Harrisburg, Pennsylvania. Before Mr. Stonebraker started his teaching career, he ran the electrical department of Saunders Electric in York. His teaching career began at York Vo-Tech in 1985, teaching Residential 1 and 2. In 1992, he began teaching commercial wiring 1 and 2 and in 1994, he began teaching Journeyman Year 1 and 2. Allen joined YTI Career Institute in 2007 as an instructor for the Electrical Technician program.

Additional contributions made by:

Fred Warner has served on several state electrical subcommittees and serves as chairperson for the Capital Region Building Code Construction Board of Review for the New York State Department of State Codes Division. He is an adjunct professor and electrical consultant and lives in Jay, New York.

Paul Westrom is an associate professor of electrical technology at the New England Institute of Technology in Warwick, Rhode Island, and a licensed journeyman electrician in the state of Rhode Island. He has a bachelor's degree in business from the New School in New York City and a master's degree in education from Capella University in Minneapolis, Minnesota. Mr. Westrom is also a retired U.S. Navy electrician's mate.

Colleen Corrice is a professional writer and developmental editor with over two decades of experience specializing in scientific and technical material. Her list of work, in the form of educational textbooks, includes trade books, test banks, and standardized tests and covers topic areas from the medical and life sciences to the automotive and electrical trades. She holds degrees in biology and scientific writing from Cornell University.

Contents

How to Use This Book

This reference guide combines the theory and direction of traditional textbooks with the educational and testing material of conventional workbooks. The result is a unique, unmatched reference that offers you the guidance you need to familiarize yourself with the type of content on the electrician's coding exam, as well as the test material you can use to experience an exam's approach firsthand.

Because standards and test questions can differ from state to state and from municipality to municipality, this book takes an unparalleled, two-pronged approach: to guide *and* to test. This approach centers on two areas: 1) general information that applies to electrical knowledge throughout the country, and 2) how to locate and reference this information in the NEC 2008 Codebook. To ensure that you gain real-world testing experience—wherever you live—questions mimic those found on actual exams. Questions may test specific information not covered in a chapter, but the underlying principles and approach have already been made clear.

Before reviewing the topics surrounding the Electrician's Journeyman and Master Licensing Exam, test your existing skills by taking the pretest at the beginning of the book. You will encounter 175 questions that may be similar to the types you will find on a state or local licensing exam. When you are finished, check your answers on page 243 carefully to assess your results. Each answer in the Answers and Explanations section provides a cross reference to the electrical code addressed by the question. Use the correlation guide in Appendix A to connect these codes to specific chapters in this book as well as in the NEC codebook. Your pretest outcome will help you determine how much preparation you need and in which areas you need the most careful review and practice.

Next, begin your review of the chapters. The content of this book is setup to align with the electrical codebook, and each chapter contains a review section so that you can further test your knowledge at your own pace.

The following special feature boxes, present in each chapter, will provide further clarification of key content areas.

Job Connection

Through the experiences of actual electricians on the job, these real-world scenarios illustrate the importance of having the right professional knowledge and experience.

ALERT!

These boxes highlight, in clear detail, the potential pitfalls facing all electricians.

Did You Know?

Helpful facts round out the book's guidance information, providing material that is both useful and interesting.

Hints and Tips
Bolstered by the experience of electricians well-versed in the trade, this feature helps new professionals navigate wiring challenges. It also points out how state and local standards can differ.

Once you have explored each chapter and feel you have increased your knowledge in the areas indicated by the pretest, go ahead and take one of the posttests found at the end of the book. Two posttests have been provided to ensure you are comfortable with the types of question you may encounter on your licensing exam and to ensure you have as much practice with your codebook as possible prior to taking your exam. We also provide an online practice test with instant scoring.

Preparing for the Exam

Whether you are just starting out in the electrical trade or are a seasoned veteran looking for career enhancement, successful completion of the electrician's licensing exam is an important step in your career. The goal of the exam is two pronged: (1) to test knowledge and (2) to determine whether applicants can apply electrical codes in real-world situations. Most states allow applicants to bring indexed copies of their codebooks to the exam. In such an open-book environment, applicants theoretically have all the answers, and the exam simply tests whether applicants know where to find them. You can set yourself up for success by familiarizing yourself with your codebook long before the exam.

▶ Familiarizing Yourself with the Exam

The exam questions you will face are based on safety, in part because the National Electrical Code (NEC) is adjunct to the National Fire Protection Association (NFPA). States may amend electrical code to reflect their individual characteristics, so questions sometimes vary according to location. In San Francisco, for example, questions on the electrical licensing exam address whether installations are earthquake safe. In Louisiana, where hurricanes are prevalent, questions address wiring suited for winds, rain, and flooding. To ensure you know the exam requirements in your area, develop a relationship with your local licensing agency.

▶ Being Ready to Meet Requirements

Like the questions on the exam, the requirements for taking the licensing exam vary depending on where you live. Some states and municipalities require that you complete an apprenticeship to take the exam. An apprenticeship is a relatively simple requirement, however. As you prepare for the exam, you work for a licensed electrician and receive on-the-job training. Because electricians are required to know electrical codes to do installations, apprentices learn codes as well as job skills. Apprentices who have been trained properly on the job will know the answer to a question asking for outlet height requirements because they have been following the relevant codes in their work. Often, aspiring electricians will take classes to prepare for the licensing exam, perhaps at a community college or trade school. In some states, class time can be applied to the apprenticeship requirement. In a few states, a practical, hands-on exam accompanies the licensing exam.

▶ Targeting the End Goal

Of course, the goal of the electrician's licensing exam is to award you a license—and render you a professional electrician. As a licensed electrician, you'll work with a number of other professionals, the electrical code examiner choice among them. Electrical code examiners have one goal: to ensure every licensed electrician provides safe and secure service. Keep this in mind as you work, and ask questions about your employer's methods. A curious mind is the best way to master your craft and help you forge sound working relationships with your code examiners. For example, if you are asked to install a refrigerator receptacle on a dedicated circuit, ask, "What's unsafe about putting a refrigerator on a general-purpose circuit?" When you ask the question, you'll learn that should a general-purpose circuit trip, perhaps from microwave-oven malfunction, the refrigerator on it would shut off and cause food spoilage.

Your clients, and your community, deserve safe, professional-grade electrical service. With sound preparation and this code guide in hand, you are on your way to wiring success.

▶ Setting Yourself Up for Success

With this book, you'll gain insight into test methodologies, reasoning, and ways to maximize your chance for success. In a sense, the NEC exam concerns itself less with what you know and more with how well you know where to find information. Many states give an open-book examination depending on local rules and regulations, therefore, you take the test with all the answers—your codebook—in your possession. Your first step toward success is to really know your codebook.

Of course, there are other steps to success. You can complete a number of administrative activities long before the test. First, contact the agency administering the test to determine which forms you must submit and by when—and then do it. Next, be sure to know what fees are involved, and pay them on time. Provide all needed information, and allow time for processing. Adhering to all agency requirements will help ensure your success.

When you speak with your agency representatives, find out what you can and cannot bring into the test. To avoid surprises, write things down. Mental preparation is key. Before you set out to take the exam, justify your need for certification and licensing. When you realize the exam is valuable, you'll be more successful. Respecting the exam and adhering to guidelines are essential steps to working safely and harmoniously in and out of your trade.

As you learn your craft, learn as much as you can about other trades—see the big picture. Seeing the big picture means remembering many details. So know yourself—know your strengths and weaknesses. The exam will challenge you at all levels. Be prepared.

To meet the exam's challenges, set goals that will help you reach success. Develop a specific timetable for goal attainment. When will you have the codebook's table of contents memorized? When will you be able to pass the practice exams with a score of 70 percent? A score of 80 percent? A score of 90 percent or better?

As you study, strive to really understand the principles underlying the questions. For example, a question may ask you to grade the receptacles in medical facilities. The answer might be, "Receptacles for inpatient sleeping beds or procedure table beds used in a critical patient care area must be listed as 'hospital grade' (NEC 517.18[B])." Notice how the answer gives both the requested information and ties that information to the codebook. That is the expectation of a tradesperson on the job: to use the codebook to provide needed information. Can you envision setting up a medical facility and having to answer this type of question to guide you through the installation? You would do exactly what you did on the exam, and that's what the exam is designed to determine. Can you—and will you—provide the information you need to act safely and responsibly as you carry out the duties of the trade? Professional electricians need not necessarily have all the answers, but they certainly need to know where to find them.

The night before you take the exam, be sure to get a good night's rest. On the way to the exam, take your time—give yourself plenty of time to arrive. Make sure you know where to take the test and when it will be given. Consider a practice run a few days beforehand.

As you launch your study of the electrical trade, know that knowledge and skills testing will always be part of your career. In addition to the other licenses and certifications you elect to pursue—all of which will require exam preparation—every day on the job will be a test. Becoming an electrician offers you an opportunity to have a long and lasting career and to acquire skills that will last a lifetime. Use this book as a review and practice guide, and start heading in the right direction. It all starts here. Good luck.

Pretest

The electrician's licensing exam is designed to assess a specific skill: whether applicants can find answers in their codebooks. To help you build speed when researching your codebook, this pretest primarily targets code references rather than theory. Use your knowledge of the National Electrical Code (NEC) table of contents and how NEC information is organized both to identify correct answers and eliminate incorrect ones.

Your licensing exam will be based on the electrical code and you should have your codebook with you when you take this pretest. Answer every question; however, if you are not sure of an answer, put a question mark by the question number to note that you are making a guess. On the official exam, an unanswered question is counted as incorrect, so making a good guess is an important skill to practice. When you are finished, check your answers on page 243 carefully to assess your results. Your pretest score will help you determine how much preparation you need and in which areas you need the most careful review and practice.

1. NEC Article ______ defines breakers and fuses.

a. 90
b. 100
c. 250
d. 500

2. Annex ______ of NEC covers conduit fill.

a. A
b. B
c. C
d. none of the above

3. NEMA stands for National Electrical Manufacturers Association.

a. true
b. false

4. What does it mean when a receptacle has "NEMA 5-20R" on it?

____
____
____
____

5. NEC Tables 310-10 through 310-21 provide information on

a. wire sizing.
b. conduit fill.
c. temperature derating.
d. none of the above

6. Which NEC table(s) cover(s) the fill of metal boxes?

____
____
____
____

7. NEC Article ______ addresses EMT.

a. 250
b. 258
c. 358
d. 900

8. Which NEC articles cover the installation requirements for RMC, IMC, and EMT?

____
____
____
____

9. [symbol] is the symbol for ______________.

10. [symbol] 3 is the symbol for ______________.

11. [symbol] is the symbol for a solar panel.

a. true
b. false

12. [symbol] is the symbol for ______________.

13. [symbol] is the symbol for ___________________.

14. [symbol] 246 represents ___________________.

15. [symbol] D is the symbol for

a. sinusoidal power source.
b. zero crossing detector.
c. dual receptacle.
d. dimmer switch.

16. In NEC 2008, what does Table 370.6A cover?

____

____

____

____

17. What NEC table would you go to to find trade size of conduit and raceway size?

____

____

____

____

18. NEC Article ______ covers metallic and non-metallic raceways in buildings.

19. The Panel Board Requirements for Application in Construction appear in NEC Article

a. 60.
b. 100.
c. 408.
d. 533.

20. The smallest kitchen counter requiring a receptacle is ______ inches.

a. 5
b. 6
c. 12
d. none of the above

21. What new code is needed for protection in all habitable spaces?

____

____

____

____

22. A 12-foot hallway requires one receptacle to meet code.

a. true
b. false

23. NEC Article 100 is dedicated to which of the following?

a. authorities
b. references
c. jurisdictions
d. definitions

24. The main ground on a service is a(n) ___________________.

25. What does NEC Section 250-28 require?

____

____

____

____

26. NEC Section 250-28A provides information on what material can be used for a main bonding jumper.

a. true
b. false

27. A bonding screw is what color?

a. blue
b. red
c. yellow
d. green

28. A subpanel requires a bonding screw.

a. true
b. false

29. Name the three methods of sizing bonding jumpers outlined in NEC Section 250-28D.

____

____

____

____

30. NEC Table ______ covers grounding electrode conductor sizing.

a. 100.12
b. 250.66
c. 300.63
d. 383.5(G)

31. Which NEC reference identifies wire sizing and the temperature a wire can handle in amps?

____

____

____

____

32. Define a grounding system according to NEC.

____

____

____

____

33. What is the topic of NEC Section 250-122?

a. grounding rod size
b. grounding conductor size
c. meter positioning
d. HVAC ductwork grounding

34. What is the NEC definition of a circuit breaker?

____

____

____

____

35. NEC Section 300-13 covers continuity conductors.

a. true
b. false

36. What does NEC Section 250 require?

____

____

____

____

37. __________ are the subject of NEC Section 680-20.

a. Portable swimming pools
b. Permanent sump pumps
c. Permanent swimming pools
d. Extension cords

38. NEC Section 250-118 covers equipment grounding conductors.

a. true
b. false

39. Which of the following NEC sections covers identification of equipment grounding conductors?

a. 100.62(B)
b. 150
c. 250.119
d. 270.2(A) through (C)

40. NEC Section 250-126 covers ground terminal identification.

a. true
b. false

41. Bonding is the establishment of __________ connection between metal parts.

42. NEC Section 250-134 provides information on what topic?

____

____

____

____

43. Equipment grounding conductor __________ is (are) discussed in NEC Section 250-130.

a. length
b. diameter
c. ampacity
d. connections

44. NEC Section 250-118 states that a grounded conductor must run __________ flexible conduit.

a. parallel to
b. perpendicular to
c. outside of
d. inside of

45. NEC Section ________ oversees bonding of flexible type conduit.

46. Which NEC article covers cable trays?

____

____

____

____

47. What is the proper procedure for bonding a motor to conduit with metal flex?

____

____

____

____

48. Threaded pipe may not be used for bonding purposes.

a. true
b. false

49. Shoulder fittings with locknuts are allowable as bonding devices.

a. true
b. false

50. What kind of information does NEC Section 250-66 cover?

____

____

____

____

51. NEC Section 250-102 is devoted to equipment

a. grounding jumpers.
b. bonding jumpers.
c. bonding terminations.
d. none of the above

52. Give guidelines for using sheet metal screws as bonding agents.

____

____

____

____

53. Which NEC article covers service drops?

____

____

____

____

54. According to NEC, trees are approved service conductors.

a. true
b. false

55. When a service entrance wire is under at least 2 in. of concrete, a disconnect is required between meter base and panel.

a. true
b. false

56. Above sidewalks accessible by pedestrians, a service drop not over 600V must be at least ________ ft. high.

a. 10
b. 12
c. 18
d. Service drops over 240V may not run over sidewalks.

57. What is the service drop distance, in feet, of a 200-amp service crossing residential property?

____

____

____

____

58. The required clearance for a 400-amp service drop across a major roadway is

a. 10 ft.
b. 12 ft.
c. 15 ft.
d. 18 ft.

59. Describe the information covered in NEC Article 230.9.

____

____

____

____

60. According to NEC 230.9(A), an unsheathed service conductor must be ________ ft. from a window that opens.

61. Which of the following NEC articles covers grounding?

a. 100
b. 150
c. 250
d. 300
e. none of the above

62. System and circuit grounding involves taking alternating current to ground.

a. true
b. false

63. The process of equipment ____________ exposes current-carrying parts to a grounding system.

64. Small power tools are protected in what two ways?

____

____

____

____

65. When bonding a plumbing system to the electrical system, it is not necessary to jumper the electric meter as this connection is built into the meter.

a. true
b. false

66. Explain the proper way to ground a subpanel in the same building.

____

____

____

____

67. Discuss the proper way to ground a subpanel in another building.

____

____

____

____

68. Something that is ____________ is connected to ground or to a conductive body that extends the grounding connection.

69. Describe a grounded conductor.

____

____

____

____

70. Define the term *grounding conductor.*

____

____

____

____

71. A(n) ____________ is the fundamental particle of electricity.

72. Voltage can be fatal to a person.

a. true

b. false

73. Which of the following determines the severity of an electrical shock?

a. voltage level

b. amount of protective clothing

c. moisture on skin or outerwear

d. all of the above

74. Grounding reduces the voltage that might cause a fatal electrical shock by allowing excess current to flow to the ground.

a. true

b. false

75. ____________ and ____________ are two types of electrical faults.

76. On a 400-amp service, how many ground rods are required?

____

____

77. The standard length for IMC is

a. 2 ft.

b. 5 ft.

c. 10 ft.

d. 12 ft.

78. Draw a properly grounded delta system.

79. Draw a properly grounded wye system.

80. NEC allows electricians to ground gas pipes to existing, grounded water pipes.

a. true

b. false

81. When is building steel permitted in an electro grounding system?

____

____

____

____

82. ____________ determines what size ground wire is used on a service.

83. A grounding conductor is what color?

a. green

b. white

c. yellow

d. black

84. What is meant by the term *isolated grounded reciprocal*?

____

____

____

____

85. Explain the process of cad welding.

____

____

____

____

86. Electricians may use rebar as a grounding means.

a. true

b. false

87. Electricians must use a(n) ____________ to properly connect number 4 copper to a ground rod.

88. Using NEC Section 250-8, identify five methods of making connections.

____

____

____

____

____

89. Underground service is simpler to troubleshoot and repair than overhead service.

a. true

b. false

90. In NEC, Table ________ provides the sizing of grounding electrode conductors.

91. Describe the process for bonding three 200-amp disconnects to make a proper grounding system.

____

____

____

____

92. In terms of AWG, what is the minimum grounding conductor size permitted?

____

____

____

____

93. The way to ground a transformer in a commercial building is to place a grounding rod inside the building.

a. true

b. false

94. Why do stoves require a four-wire system?

____

____

____

____

95. Why do dryers require a four-wire system?

____

____

____

____

96. Which of the following do not require a four-wire system?

a. air-conditioning units

b. hot water heaters

c. baseboard heat

d. all of the above

97. Which of the following NEC sections outlines which air-conditioning systems require no grounding?

a. 110.73

b. 156.13

c. 250.114

d. 255.74

98. Explain where and why isolated systems are used.

____

____

____

____

99. Electricians should put grounded conductors on yellow reciprocal screws.

a. true

b. false

100. On a 200-amp service, the electrician should use ______________ or ______________ size grounded conductor in the service cable.

101. When using 500 MCM, ______________ MCM or ______________ MCM size ground wire is appropriate on a 400-amp service.

102. Name a common application for high impedance grounding.

____

____

____

____

103. What is the proper way to test conduit for proper grounding?

____

____

____

____

104. How does an NEC violation affect the occupancy of an existing installation?

____

____

____

____

105. The maximum operating temperature of type FFH-2 fixture wiring is

a. 30°F.

b. 30°C.

c. 75°F.

d. 75°C.

106. NEC Article ________ addresses bonding requirements in healthcare facilities.

a. 220

b. 300

c. 517

d. 780

107. FMT conductors are appropriate for battery storage areas.

a. true

b. false

108. A circuit that cannot cause a fire in a hazardous condition is called a(n) ______________ ______________ system.

a. hermetically sealed

b. intrinsically safe

c. closed loop

d. open loop

109. When local ordinance supersedes or administers NEC, NEC uses what phrase?

____

____

____

____

110. A(n) ____________ circuit is designed and created for one load.

a. specialized
b. industrial
c. dedicated
d. time-delay

111. NEC Article ____________ specifies conduit fill requirements for fiber-optic cabling.

a. 380
b. 600
c. Chapter 9 tables
d. 821

112. Chapter 3 of the NEC discusses fixture wiring and fixture wires.

a. true
b. false

113. NEC 90.1(A) states that the purpose of the NEC is the "practical safeguarding of ____________ from hazards arising from the use of electricity."

a. people
b. buildings
c. property
d. pets
e. businesses
f. all of the above

114. NEC Article 426 recognizes which three kinds of heating systems?

____

____

____

____

115. What is proper phase arrangement when wiring a three-phase panelboard?

____

____

____

____

116. When installing an autotransformer, it is important to remember to insert an OCPD in series with the shunt winding.

a. true
b. false

117. NEC guidelines guarantee the best workmanship and safest installations possible.

a. true
b. false

118. When NEC Article 100 states, "All circuit conductors between the service equipment, the source of a separately derived system, or other power supply source and the final branch-circuit overcurrent device," what is it referencing?

____

____

____

____

119. NEC Article ________ outlines the restrictions for using plastic junction or device boxes.

120. NEC Chapter 5, which concerns hazardous locations, ______________ NEC Chapters 1–4.

a. supersedes
b. supplements
c. modifies parts of
d. has no relationship to

121. Per NEC 503.4, circuit breakers in Class III hazardous environments require which type of enclosure?

a. hospital-grade
b. childproof
c. waterproof
d. dust-tight

122. General-purpose enclosures are NEC approved for intrinsically safe apparatuses.

a. true
b. false

123. Fusing the load side of a transformer sufficiently protects the transformer.

a. true
b. false

124. The NEC definition "the current, in amperes, that a conductor can carry continuously under the conditions of use without exceeding its temperature rating" describes which of the following?

a. power rating
b. ampacity
c. temperature rated current
d. continuous load current

125. Where does NEC list the types of insulation material deemed acceptable in wet locations?

____

____

____

____

126. Leads to an unused current transformer should be

a. cut off and terminated close to the transformer body.
b. shorted together.
c. left open.
d. tucked into or under the transformer frame.

127. In what sections does NEC outline the requirements for appliance calculations?

____

____

____

____

128. Typically, transformers are available in one of two types: ____________ and ____________.

129. Given the wiring schedule, how is a proper raceway chosen?

____

____

____

____

130. The "3-ft. rule" for clearances around things like control panels is a
- **a.** minimum clearance that must be equaled or exceeded.
- **b.** maximum clearance that must not be exceeded.
- **c.** a and b
- **d.** suggestion, not a requirement.

131. When establishing ground clearance in new construction, it is best to make the ground clearance slightly more than the NEC-stated requirement to provide a margin for error.
- **a.** true
- **b.** false

132. Service drops must be installed so as to be "readily accessible."
- **a.** true
- **b.** false

133. Extension cords are acceptable in wet areas.
- **a.** true
- **b.** false

134. The requirements for liquid-filled transformers are the same regardless of the liquid used in the transformer.
- **a.** true
- **b.** false

135. NEC Article 526 helps electricians determine requirements for
- **a.** embedded deicing systems.
- **b.** open or exposed deicing systems.
- **c.** a and b
- **d.** none of the above

136. Disconnects and switches used with appliances must indicate, on their switches, the on- and off-positions.
- **a.** true
- **b.** false

137. In a three-phase delta system, the leg with the highest voltage must be
- **a.** reduced to within 1% of the other legs' voltages.
- **b.** considered the "B" leg.
- **c.** labeled using permanent marking as "HV."
- **d.** connected directly to neutral.

138. Branch circuits must be loaded equally in terms of wattage per unit area.
- **a.** true
- **b.** false

139. An electrician can safely recess a panel board into a flammable wall.
- **a.** true
- **b.** false

140. All metallic boxes and components must be grounded unless a grounding conductor is absent.
- **a.** true
- **b.** false

141. Nameplates that rate protective devices describe which of the following?
- **a.** how much current an appliance draws
- **b.** what size breaker to use to protect the circuit
- **c.** a and b
- **d.** none of the above

142. NEC maintains no ampacity requirements for fixture wiring because all fixtures undergo the UL approval process.
- **a.** true
- **b.** false

143. The calculations for protecting motor circuits are the same as those for nonmotor circuits.
- **a.** true
- **b.** false

144. Per NEC Article 410, luminaires may mount on drop-ceiling tiles when those luminaires
- **a.** weigh less than 5 pounds.
- **b.** are less than 3 in. in diameter.
- **c.** reside in a dry area.
- **d.** none of the above

145. The requirements for burying 240V conductors are found in NEC ________.

146. NEC ________ outlines the requirements for AC type (armored) cable.

147. A garage door opener is considered an appliance for NEC purposes.
- **a.** true
- **b.** false

148. The maximum motor horsepower that can be disconnected using a panel breaker is

____
- **a.** 0.1.
- **b.** 0.125.
- **c.** 0.25.
- **d.** 1.

149. When running new work over old, hanging a second support from a primary support is an acceptable practice.
- **a.** true
- **b.** false

150. Installation of a floor drain in a vault intended for wet transformer(s) is
- **a.** required.
- **b.** required when the vault has more than 100KW of transformer capacity.
- **c.** good practice regardless of rules or requirements.
- **d.** none of the above

151. A TVSS provides adequate protection against lightning strikes.
- **a.** true
- **b.** false

152. It is possible to violate the NEC by tightening a bolt or fastener too tight.
- **a.** true
- **b.** false

153. When installing a cable tray, only the sections nearest the terminations must be grounded.
- **a.** true
- **b.** false

154. Using aluminum wire with receptacles is
- **a.** against code.
- **b.** acceptable only in uninhabited residential areas.
- **c.** appropriate only in certain commercial applications.
- **d.** safe when the receptacle is correctly rated and listed for such use.

155. How would demand be calculated in a commercial environment in which a branch circuit could effectively handle 20 KW?

____

____

____

____

156. NEC allows how many disconnects per service entrance?
- **a.** one
- **b.** two
- **c.** six
- **d.** unlimited

157. In a given NEC section, Subsection (2) contains
- **a.** references.
- **b.** FPNs.
- **c.** terms and definitions.
- **d.** disclaimers.

158. Overcurrent insufficient to trip a breaker can prove hazardous under the right conditions.
- **a.** true
- **b.** false

159. Which of the following requires no AFCI protection?
- **a.** branch circuit in which the initial run is encased in AC up to the first receptacle, which is mounted in a plastic box
- **b.** branch circuit in which the initial run is encased in AC up to the first receptacle, which is mounted in a metal box
- **c.** clothes closet measuring more than 10 square ft.
- **d.** none of the above

160. In a multiphase system in which one phase is used as a single-phase source, which conductor serves as ground?

____

____

____

____

161. In NEC terms, what does the phrase *in sight* mean?

____

____

____

____

162. What is the difference between NFPA 70 and the NEC?
- **a.** NEC is part of NFPA70.
- **b.** NFPA70 is part of the NEC.
- **c.** NFPA concerns itself with fire safety, the NEC concerns itself with electrical safety.
- **d.** There is no difference.

163. Property owners can act as their own AHJs.
- **a.** true
- **b.** false

164. What demand factor applies when calculating loading requirements for an apartment building with 21 units? A building with over 100 units?

____

____

____

____

165. When considering an effective grounding path, which of the following statement(s) is/are true?
- **a.** Bonding shall be provided around insulated joints.
- **b.** Bonding shall be provided around any equipment likely to be disconnected for repairs or replacement.
- **c.** Bonding jumpers shall be of sufficient length to permit removal of equipment while retaining the integrity of the grounding path.
- **d.** all of the above

166. In NEC terms, "The permanent joining of metallic parts to form an electrically conductive path that ensures electrical continuity and the capacity to conduct safely any current likely to be imposed," defines __________________________.

167. NEC Table 250.66 applies to the derived conductors of separately derived AC systems.
- **a.** true
- **b.** false

168. A ground ring encircling a building or structure, in direct contact with the earth consisting of at least __________ ft. of bare copper not smaller than ________ AWG, shall be permitted as an electrode.
- **a.** 10, # 4
- **b.** 20, # 4
- **c.** 25, # 4/0
- **d.** 20, # 2

169. Where a main bonding jumper or a system bonding jumper is a screw only, the screw shall be identified with a ______________ finish that shall be visible with the screw installed.
- **a.** green with yellow-striped
- **b.** galvanized
- **c.** copper-plated
- **d.** green
- **e.** all of the above

170. NEC Table 220.42 can be used to calculate lighting loads in a restaurant dining area.
- **a.** true
- **b.** false

171. Outdoor lighting installations exceeding ______________ are subject to the clearance requirements outlined in NEC 225.7.
- **a.** 30 amperes
- **b.** 1,000 watts
- **c.** 120 volts
- **d.** 240 volts

172. Open conductors belonging to separate circuits must be separated by

a. 1 in.
b. 50 mm.
c. 4 in.
d. 200 mm.

173. NEC requires a panelboard circuit directory.

a. true
b. false

174. A tree over 50 ft. high may be used as an overhead service support for a single-story residence only.

a. true
b. false

175. What is the maximum number of sets of service conductors a service drop or lateral may supply?

a. one
b. two
c. six
d. no limit

Introduction: Basic Electrical Concepts

CHAPTER SUMMARY

The science of electricity is governed by the laws of physics and chemistry. The electrician's trade is governed by NEC chapters, articles, and sections, as well as by myriad local and regional requirements. With such regulations as a guide, electricians who understand how electricity behaves, how it can be converted, and how it can be harnessed can safely execute installations using the spectrum of tools designed for personal and overall protection. In whatever form, electricity is a profoundly powerful entity that inherently carries a sizeable risk. Vigilance to safety and electrical basics are the two essential elements to an electrician's success.

Learning Objectives

- Define common electrical terms
- Identify and describe the main components of electrical circuitry
- Explain Ohm's law as it relates to electrical circuitry components
- Discuss circuit efficiency in terms of loss
- Review wire gauge and ampacity
- Describe transformer action
- Outline the principles behind Watt's law

Key Terms

ampacity
ampere
circuit breaker
conductor
current
electricity
electromotive force (EMF)
fuse
gauge
horsepower
insulator
load
ohm
power
resistance
volt
voltage
watt
wattage

Introducing Electricity Basics

At the heart of the electrical profession lie a few basic concepts. This chapter explains what **electricity** is, how it moves through wires, and how it can be converted to other forms of energy that do work for us.

In general, the term *electricity* is defined as an imbalance in negative charges, or electrons, between two points. To achieve balance, the electrons on the electron-rich (negative) side feel pressure to move to the electron-poor (positive) side. This pressure to move is called **voltage**, or **electromotive force (EMF)**.

Understanding Voltage

Voltage strengthens as a negative terminal becomes crowded with electrons. Like people, electrons move from long lines to short ones. The greater the difference between two lines, or terminals, the greater the desire to move between them. Assume, for example, that the negative terminal on a battery has a surplus of electrons, produced by a chemical (battery) or physical (turbine) reaction, while the positive terminal has a corresponding surplus of protons, or positive charges. To try to achieve balance, the negative charges of the battery try to move to the positive side. The more electrons that are produced by the mechanical or chemical process, the stronger their urge to move and the greater the pressure. The greater the pressure, the higher the voltage.

Once two terminals have the same number of electrons, voltage becomes 0 and electrons stop moving.

Defining Current

Current is the flow of electrons from one place to another. Always negative to positive, current is measured in **amperes**, or amps.

Electrons follow the path that is easiest, but barriers can prevent movement. When positive and negative points are separated by a barrier or an **insulator**, like air, and voltage increases enough, the electrons will jump across the insulator to the positive terminal. This phenomenon, called arcing, can occur through air, paper, rubber, or any other insulator. When it is too difficult for electrons to pass through a **conductor** and/or **resistance**, they may take an alternate, undesired, and potentially hazardous path. Fires and other unsafe conditions may result.

To harness the potential energy of electrons' moves, electrons can be forced to do some kind of work on the path from negative to positive, much like a river can be forced to turn a turbine as it moves from high to low. In the electrical world, we might place a thin filament in the electrons' path and force all electrons to squeeze through it. In the process of traversing the filament, the electrons would produce heat and the filament would glow. This is the concept behind a lightbulb.

When electrons are idle, meaning there is no pressure or voltage, no work is done.

Did You Know?

At the dawn of the discovery of electrical power and its use, current was a big unknown because it was unmeasurable. Early scientists noted that when current to a lightbulb increased, the lightbulb burned brighter. As a result, the term *intensity*, represented by the letter *I*, was used to represent the increase or decrease in brightness based on a similar change in what we now know as current flow.

Current used to be called Intensity, thats why "I" is used in Ohms law for current.

Applying Resistance

To be useful, electrons' flow from negative to positive must be controlled with some kind of resistance. Resistance, which is measured in **ohms**, is defined as "opposition to current flow." Consider a common garden hose as an example. The restriction of the hose's nozzle increases pressure through the hose. This increased pressure increases velocity of water exiting the nozzle.

In electricity, resistance is generally used to convert one form of energy to another. As the electrons rub against each other and the walls of a filament, they create heat from friction, which causes the filament to glow. Light is the desired effect; the heat is an unused by-product.

With no resistance, all electrons travel immediately and expend all energy. The electrons' path will either burn up, as in a lightbulb, or all electrons will complete their trip on the path and there will be no more pressure, or voltage. Current that flows without resistance, or load, results in a short circuit, which is undesirable in most cases. Short circuits, like arcing, can start fires.

Outlining Circuits

A circuit is the combination of an electron source, a pathway (conductor), a **load** (resistance), and an electron sink (ground). All these components must be present for a circuit to function properly. When any part is missing, an open circuit results. In an open circuit, no current flows. In such a situation, a lightbulb, for example, will not emit light.

▶ Explaining Ohm's Law

To measure current, a device can be inserted into a circuit to count electrons as they pass by. Because this is generally impractical, electricians instead determine current using Ohm's law, expressed as:

Current = voltage/resistance

Or:

$$I = V/R$$

Figure I.1 illustrates Ohm's law using plumbing. Water, representing current, flows from a pipe (source), through a valve, and into a sink, thereby completing the circuit to ground. When the valve is opened, the pressure on the input side remains the same as the flow increases. In Ohm's law terms, current increases because resistance decreases.

Changing the diameter of the valve adjusts the flow, or current. A smaller diameter increases resistance to flow. Similarly, the diameter of a wire determines how much current the wire can carry. Wire comes in **gauges**. The lower the gauge number, the thicker the wire and the more current it can carry safely.

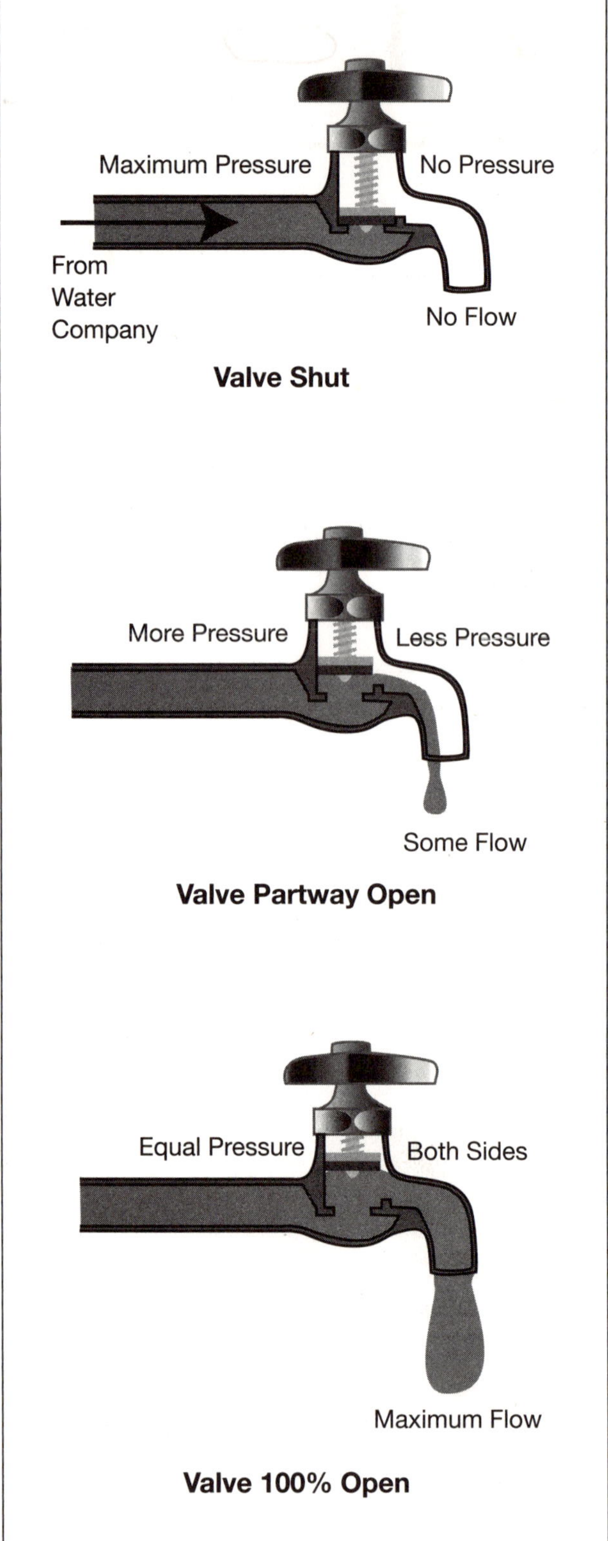

Figure I.1 Ohm's Law in Plumbing

▶ Discussing Circuit Efficiency

Electricity is a means to an end. We do not use electrical energy directly. Instead, we convert it to other types of energy and harness that energy for our use. We usually convert electricity to one of the following:

- Light—Lightbulbs account for a large portion of electrical usage. An incandescent bulb burns a filament that glows. A fluorescent

Did You Know?

Motors and generators work the same way. A motor converts electricity to rotational motion. A generator converts rotary motion to electricity. If you were to take a simple motor and turn the shaft by hand, voltage would develop across the input terminals.

bulb uses a gas that glows when excited by electrons.

- Heat—Ovens, stoves, and electric baseboard heating systems all transfer heat as energy.
- Magnetism—Magnetic fields that use electricity turn motors.

Loss occurs when energy is converted. For example, a 1-**horsepower** (746-watt) motor may require an input of 1,000 **watts**. The remaining 234 watts (1,000–746) are lost in the inefficiencies of the materials used to make the circuit. A 100-watt lightbulb may require 125 watts of electrical power to emit 100 watts of light. There are losses in the typical electrical circuit's wiring, connections (splices), and loads.

Wiring or line loss is due to resistance in the conductor. In a perfect world, conductors (copper wires) have 0 ohms of resistance. In reality, there is some slight resistance (measured per foot) that adds up over long distances. The longer a garden hose, for example, the weaker the flow at its end. Over long distances, pressure builds due to the friction of water molecules trying to squeeze through. A wider hose, one with a larger diameter, will counteract this phenomenon. This is true with electrical wiring as well. When line loss becomes a factor, the solution is to increase the wire's diameter.

▶ Illustrating Transformer Action

The power manufactured from fossil fuels and/or water is shipped through a grid of high-tension wires, each carrying thousands, or tens of thousands, of **volts**. This voltage is somehow converted to standard 110V residential voltage. A transformer makes this conversion, trading voltage for current, lowering the voltage and increasing the available current. A transformer may have, at its input, 100,000V at 1 milliamp (1/1000 of an amp) and produce at its output 100V at one amp. Note that the power at both sides (voltage × current) is the same: 100 watts. In this case, the transformer reduces the voltage by a factor of 1,000, so it would be considered a "1000:1 transformer." Note that transformers are useful only in alternating current applications (see following section).

▶ Studying Fundamental Circuits

In direct current (DC), electrons move in one direction only. Modern installations use alternating current (AC), which reverses the positive and negative terminals periodically, about 60 times a second, to keep electrons traveling between terminals. DC derives from AC using a DC power supply that essentially smoothes the ripples in the AC waveform, much like a mason smoothes cement with a trowel.

DC and AC achieve the same goal, just in different ways. Consider DC a circular saw in which the blade moves in only one direction, and AC a reciprocating saw with a blade that moves back and forth. AC is popular when large energy quantities are required, such as those sufficient to supply a house or even a city, because it is easier to make and transport over long distances.

Job Connection

Dean had a 9:00 A.M. appointment, his first, at Mrs. Cheever's house. Wearing a pair of old jeans and a ripped T-shirt and holding a few tools in his hand, he rang the doorbell promptly, but Mrs. Cheever was hesitant to let him in. Although Dean was a skilled professional, his appearance both gave the wrong impression and jeopardized his safety.

In addition to being professional and appropriate, an electrician's clothing should exceed the requirements for safe electrical work. The shoes, gloves, and protective eyewear that are designed for electrical work are rated for the volts they can withstand before allowing electrons to cross their barriers. The work gloves sold at consumer outlets, in contrast, do little to safeguard their wearers from injury.

Like electrical clothing, electrical tools must be safe. A screwdriver with a thin rubber coating may guard against residential voltages (110VAC) but offer little if any protection at 48,000V. To maintain the necessary standards, electricians like Dean should periodically test their tools and their protective gear.

Consider a circuit with two lightbulbs, a source, a switch, and a fuse. The switch opens and closes the circuit, and the fuse opens the circuit when current is excessive, as in a short circuit. The wires, called conductors, are considered to have 0 ohms of resistance to ensure both that all current reaches the load(s) and that the insulation covering the wires keeps all electrons safely in the wires where they belong.

Now assume that two lightbulbs, each with a resistance of 100 ohms, are hooked up end to end, one after the other, in series (Figure I.2A). The electrons must pass through 200 ohms of resistance, so with a 100V source the current is $\frac{1}{2}$ amp. Using Ohm's law, each lightbulb receives only 50V and is therefore only half as bright as would be expected.

If the two lightbulbs were on separate paths, arranged in parallel, each electron would receive only 100 ohms, so the current would be 1 ampere through each lightbulb (Figure I.2B). These lightbulbs would seem brighter, because they would receive the full 100V. This wiring scheme, used in household and commercial installations, is the type of circuit discussed in the balance of this book. Lightbulbs could be added, each as bright as the others.

As mentioned earlier, Ohm's law defines current as voltage divided by resistance. Voltage, therefore, is the product of current and resistance ($V = I \times R$). Resistance is voltage divided by current ($R = V/I$).

In the circuit in Figure I.3, a source provides 100V of energy and each lightbulb is directly connected to, or across, the source, thereby receiving 100V of energy. Each lamp has a resistance of 100 ohms. Therefore, the current through each lamp is V/R, or 100V/100 ohms, which results in a current of 1 ampere per lamp. The total current in the circuit is 2 amperes. This is a very tidy arrangement: all current leaving the source reenters the source. All current entering a junction leaves that junction. Consider the effect of opening a switch and turning off one of the lightbulbs (Figure I.4).

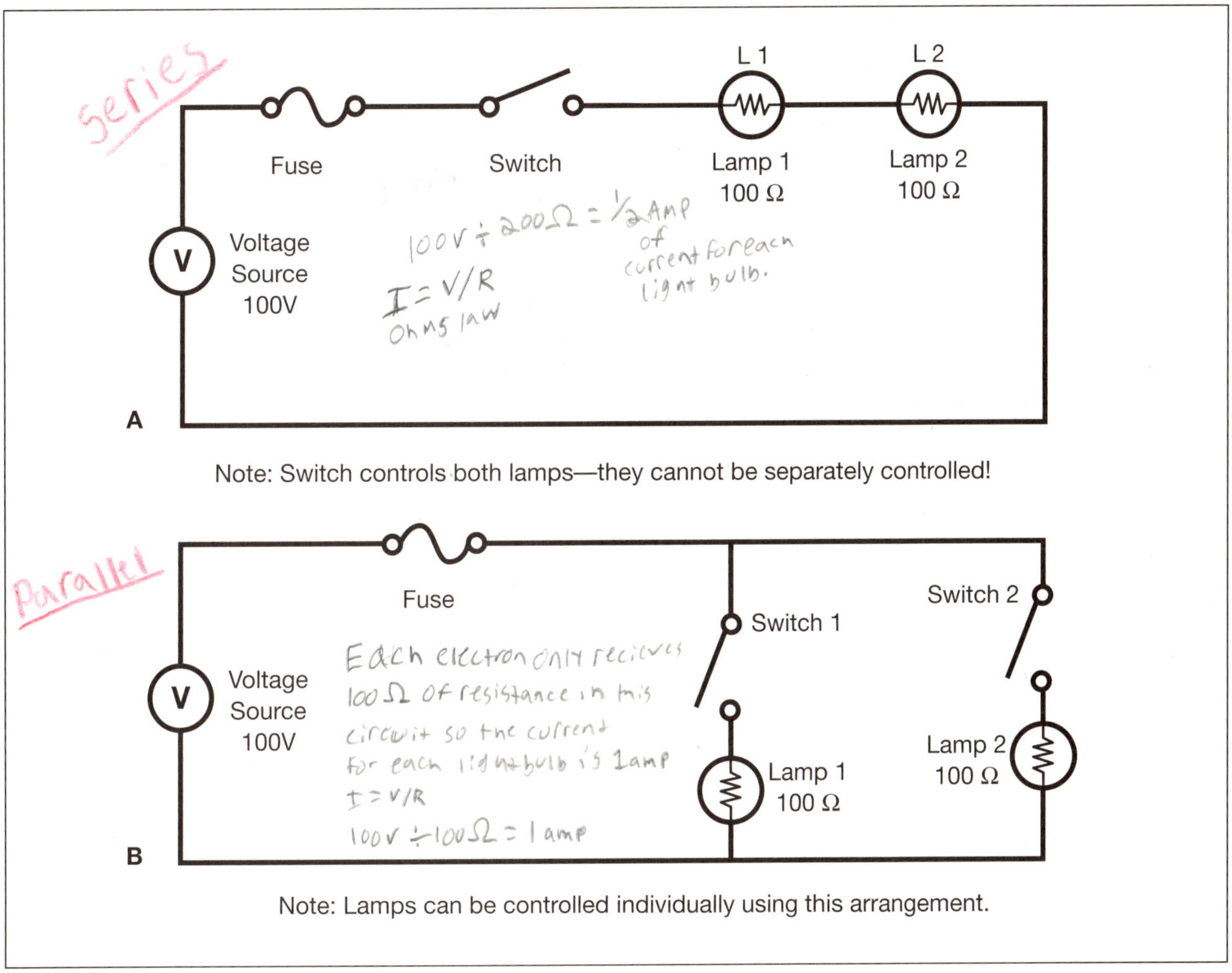

Series
L 1
L 2
Fuse
Switch
Lamp 1
100 Ω
Lamp 2
100 Ω
V
Voltage
Source
100V
100V ÷ 200Ω = ½ Amp of current for each light bulb.
I = V/R
Ohms law
A
Note: Switch controls both lamps—they cannot be separately controlled!
Parallel
Fuse
Switch 1
Switch 2
V
Voltage
Source
100V
Each electron only recieves 100 Ω of resistance in this circuit so the current for each light bulb is 1 amp
I = V/R
100V ÷ 100Ω = 1 amp
Lamp 1
100 Ω
Lamp 2
100 Ω
B
Note: Lamps can be controlled individually using this arrangement.

Figure I.2 Serial and Parallel Loading

2 Amperes
1 Ampere
1 Ampere
V
Voltage
Source
100V
Lamp 1
Lamp 2
2 Amperes
1 Ampere
1 Ampere
Note: All currents entering a point equal those leaving that point.

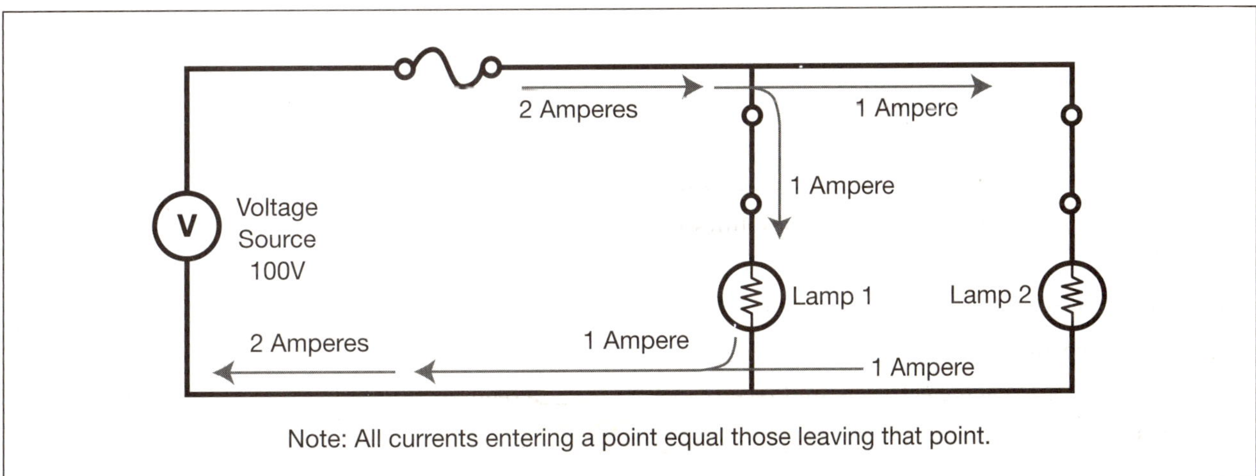

Figure I.3 Conservation of Energy

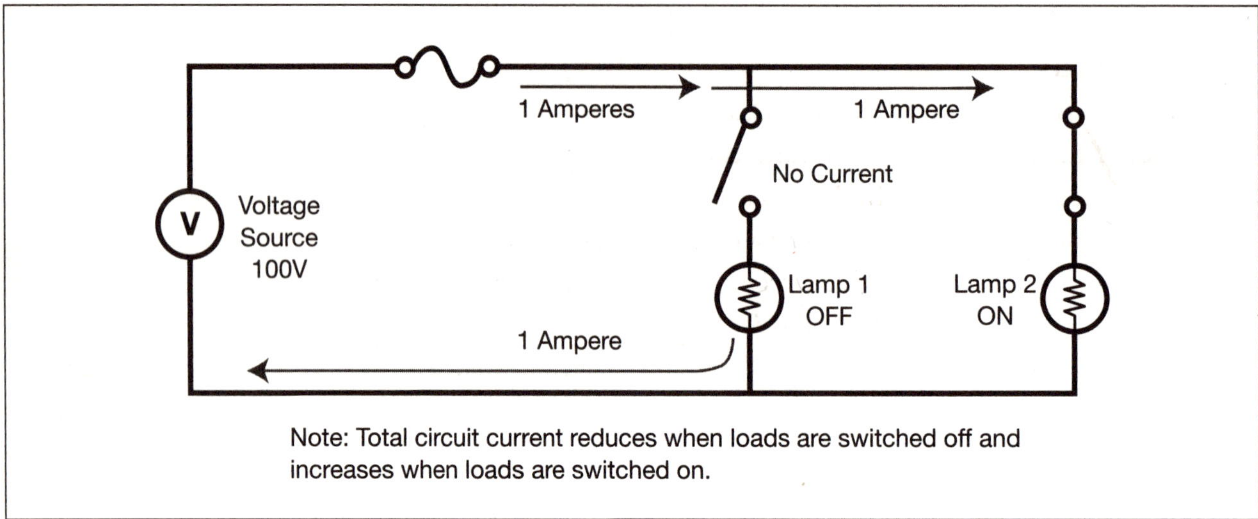

Figure I.4 Load Switching on Circuit Current

In a residence, a circuit may have several switched loads. A bedside lamp, an air conditioner, a stove, and a dishwasher are all loads that can be brought into a circuit through switching. A circuit cannot sustain limitless loads, however. Just as a lightbulb will get hot and glow with excessive current, the wiring in a residence will heat when wiring has too many loads. A **fuse**, or **circuit breaker**, is designed to burn and create an open circuit, stopping all current to all loads in the circuit. The fuse in Figure I.5 is 10 amps. When current in the circuit exceeds 10 amps, the fuse will stop all current.

Given that the fuse in this example is 10 amps and each lightbulb draws 1 amp, our circuit can handle 10 lightbulbs. An eleventh bulb would boost the circuit's

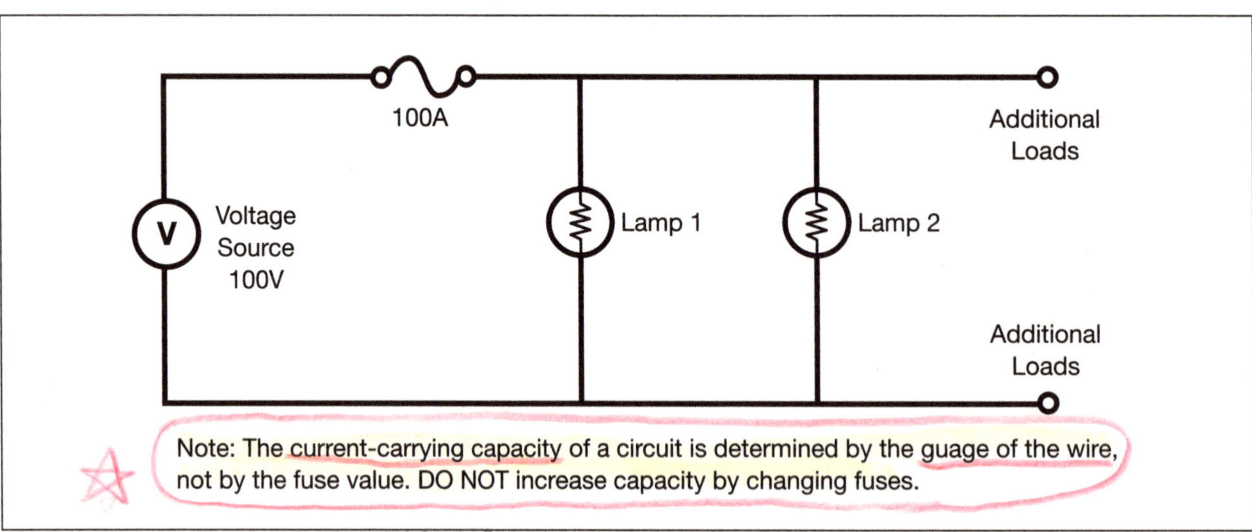

Figure I.5 Fuse Value Caution

Ampacity = current carrying capacity?

total load to 11 amps and melt the fuse. The simplest solution might seem to replace the fuse with one of a higher current rating, but doing so would exceed the capabilities of the circuit wiring and other components. Heat would become excessive, leading to circuit failure at best and fire at worst. Increasing a circuit's **ampacity** requires either replacing the wiring with a larger gauge or developing a separate circuit. This is why fuse boxes and breaker panels have many fuses and/or breakers.

▶ Learning Watt's Law

All the preceding calculations are based on Ohm's law. Lightbulbs are rated not in ohms but in watts, a unit of **power.** Referring once again to Figure I.3, we see that each lamp consumes 100W of power (1 ampere × 100V) for a total circuit power of 200W. Generally, consumers are concerned with **wattage**, because electricity is sold by the watt. Electricians, however, are concerned more with amps and volts because these are the physical NEC focuses on currents, because voltages are standard and current is the electrical characteristic that poses the biggest safety threat.

To help ensure safety, the NEC sets out detailed guidelines for circuit layout and installation. Like most residential and commercial circuits, the circuits examined so far are very simple. Making circuits consistently reliable and safe is difficult. Circuits are rated by their current capacities and require specific components to deliver amperage safely. A 10-amp circuit, for example, requires a 10-amp fuse or breaker, at least 14-gauge wire, and receptacles and switches rated to carry that much current. A 15-amp circuit requires different, more robust components; a 20-amp circuit, even more so.

In a circuit, wire insulation is color-coded according to function. Black wire, for example, is the "hot" wire that acts as the source. Once it passes through the load, a white wire carries that current back to the source. All voltages in a circuit are referenced to Earth ground (0V) through a green wire, which may or may not accompany each receptacle or load.

Wire labeled "14-2" has two conductors, one black and one white, each of 14-gauge diameter. Wire labeled "14-3" has three conductors: one white, one black, and one green ground wire. Wire labeled "12-2" and "12-3" is similar except that it is thicker and can handle more current (Figure I.6).

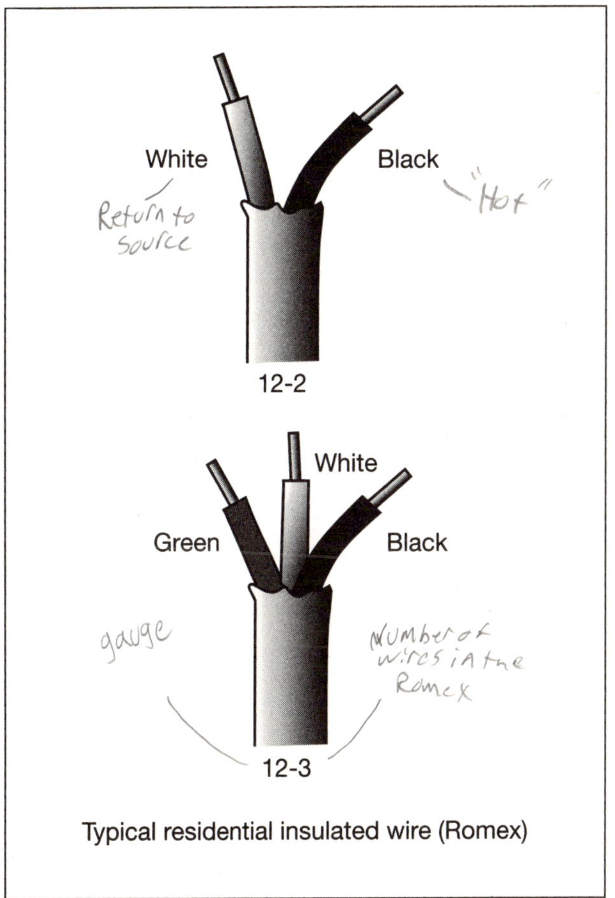

Figure I.6 Grounded and Ungrounded Residential Wire

Figure I.7 Circuit Switches Installed and Ready for Finish

In addition to delivering safety and security, circuit connections must add no significant resistance to a circuit. Typical connection methodologies include soldering and wire nut splicing. Simply taping wires together after twisting is unacceptable. All splices must reside inside a sealed junction box but remain accessible for inspection, upgrade, and repair (Figure I.7).

In commercial settings, conductors reside inside tubing, called conduit, which provides isolation as well as an extra level of protection not afforded by plastic coating or sheathing (Figure I.8). Such measures are not always necessary in residential applications.

Figure I.8 Commercial Conduit Installation

Figure I.9 Typical Circuit Breaker

The predominant way to protect circuits and their loads is circuit breakers (Figure I.9), although fuses can provide overcurrent protection. Circuit breakers trip when current exceeds the breakers' rated values. Unless it is damaged, a breaker can be reset to function when the cause of the overcurrent is identified and removed. Figure I.10 shows a clamp-style multimeter that is used to verify voltages, currents, and resistances in a circuit. The clamp is really just a loop that fits around a conductor, thereby eliminating the need to break the circuit to measure the current in the conductor.

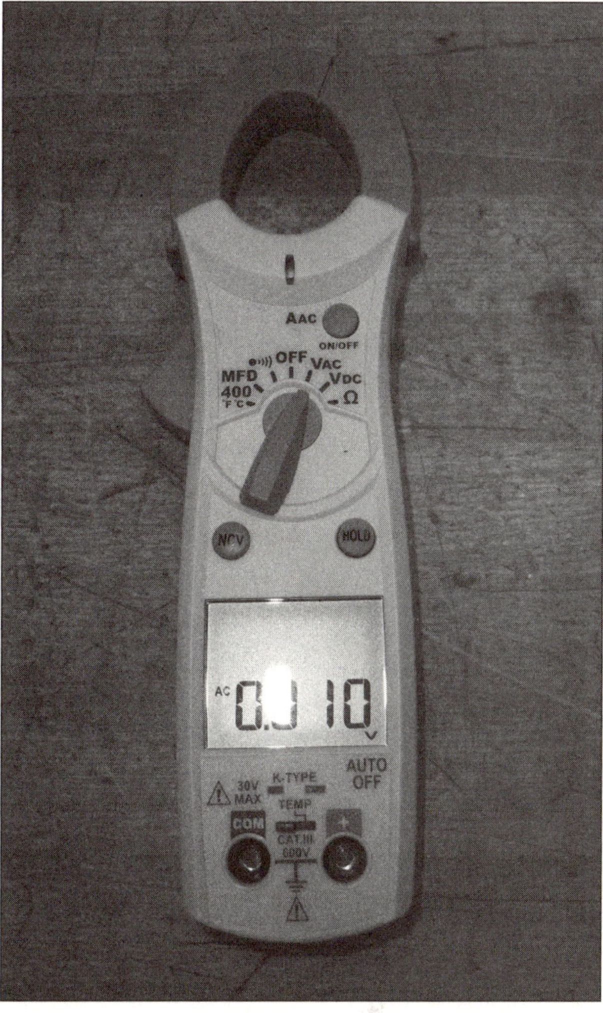

Figure I.10 Typical Modern Clamp-Style Multimeter

▶Chapter Review

1. In electric circuits, a load usually converts electrical energy to ______________, ______________, or ______________.

2. A circuit has a source of 10V and a resistance of 2 ohms. A fuse in this circuit should have what value?

____

____

____

____

3. In electrical work, electrons are always balanced between two points.

a. true

b. false

4. Differentiate between AC and DC.

____

____

____

____

5. Which of the following is not a conductor?

a. gold

b. silver

c. copper

d. water

e. air

6. Current is measured in ______________.

7. Insulators like air, paper, or rubber prevent arcing.

a. true

b. false

8. A circuit using 2 amps of current with 12V applied dissipates how much current?

____

____

____

____

9. All electrical equipment and supplies should be considered ______________ until determined otherwise.

a. active or live

b. shorted

c. dead or off

d. open

10. Voltages of ______________ or more are considered high.

a. 480

b. 240

c. 208

d. 120

11. A voltage of 1V applied across a 1-ohm resistance produces a 1-amp current. If the voltage is doubled and the resistance is halved, the current does which of the following?

a. increases four times

b. doubles

c. remains the same

d. increases three times

12. All voltages in a circuit are referenced to Earth ground via a black wire.
- **a.** true
- **b.** false

13. Which of the following best defines a parallel circuit?
- **a.** All current must flow through all loads.
- **b.** Each load is connected in a separate path.
- **c.** A parallel circuit is a combination circuit.
- **d.** All loads are connected one after the other.

14. The most important aspect of becoming an electrician is learning about ______________.

15. A transformer has a voltage ratio of 8:1 and an input of 120V. What is the output voltage?

____

____

____

____

16. When an ammeter is connected in parallel with a load,
- **a.** no reading is indicated.
- **b.** damage to the meter and/or circuit is probable.
- **c.** it reads current as expected.
- **d.** voltage is the only reading.

17. Motors and generators work the same way.
- **a.** true
- **b.** false

18. The organization that sponsors the NEC is the
- **a.** American Society for Testing of Materials (ASTM).
- **b.** National Fire Protection Association (NFPA).
- **c.** Electrical Safety Commission (ESC).
- **d.** National Electrical Manufacturers Association (NEMA).

19. The negative terminal of a battery has a surplus of protons.
- **a.** true
- **b.** false

20. Which of the following lacks insulation properties?
- **a.** rubber
- **b.** air
- **c.** water
- **d.** mica
- **e.** paper

21. Current always flows from negative to positive.
- **a.** true
- **b.** false

22. In the standard U.S. system of wire sizes, from No. 50 through No. 4/0, the lower the number, the ______________ the wire.
- **a.** smaller
- **b.** heavier
- **c.** longer
- **d.** thicker

23. The major cause of death from electrical shock is

a. head trauma.
b. lack of oxygen.
c. burns.
d. heart fibrillation.

24. As the diameter of wire increases, resistance to flow increases.

a. true
b. false

25. A(n) ____________ acts as a fuse, but is usually resettable.

26. If you have 750 watts total in bulbs, how many amps is it at 120V?

a. 8.4
b. 7.3
c. 6.25

27. Resistance is generally used to convert energy from one form to another.

a. true
b. false

28. The larger the diameter of a piece of wire, the ____________ its AWG value.

29. If you have 1,800 watts for baseboard heat, at 240V how many amps are there?

a. 6
b. 7.5
c. 10

30. Voltage strengthens as a negative terminal becomes crowded with electrons.

a. true
b. false

PART

I

In the Home: Residential Electricity

CHAPTER

General Residential Requirements

CHAPTER SUMMARY

When installed properly and according to requirements, the main elements of a residential electrical installation—things like the panel, junction boxes, switches, and receptacles—come together to deliver service that keeps homeowners safe and illuminated and their appliances running. Branch circuitry is at the heart of residential wiring and the benchmark for measuring the successful electrician's job competency. In addition to being able to identify branch circuit types, residential electricians must be adept at executing the circuit methodology that ensures, in simplistic terms, that wire, breaker, device, and load ampacities all work in harmony. The goal, overall, is simply safety. Beyond merely replacing fuses and circuit breakers correctly, the successful electrician identifies and implements appropriate protective measures at all levels. Tasks like calculating box fill and grounding installations properly are merely means to that goal.

▶ Learning Objectives

- Discuss the term *residential electrical service*
- Define the term *branch circuit*
- State branch circuit requirements
- Identify protection levels and methods for residences
- Calculate box size based on device and conductor population and conductor gauge
- Explain how to ground a residential installation properly

▶ Key Terms

branch circuit
fuse
ground fault
panel
receptacle
Romex
service entrance
subpanel

▶ Understanding the Parts of Residential Installations

While the National Electrical Code (NEC) fails to differentiate between residential and commercial services, the two exhibit one main difference: at times, all people in residential environments are asleep, while at least one person is always awake and in charge in a commercial one.

Did You Know?

Although people sleep in them, for electrical purposes apartment buildings and condominiums are considered residential/commercial hybrids.

Most residential installations have the following main electrical parts:

- Service entrance/meter
- Panel, fuse box, load center, or breaker box
- **Romex** wire, 6, 8, 12, or 14 Ga.
- Junction box(es)
- Switches
- Receptacles
- Appliances and fixtures
- Splices

This chapter explains the general considerations of residential electrical parts. For details, see Chapter 3.

Working with the Service Apparatus

Electrical services are rated by supplied amperes (e.g., 100 or 200). Residential service equipment most commonly comes in 100-, 150-, and 200-ampere sizes. In residential installations, service amperage and residential service equipment must be equal to or greater than the load being applied (NEC 310.15[B][6]). Table 1.1 is a reference tool for this purpose (NEC Table 310.15[B][6]).

In addition to matching the applied load, residential service equipment must be grounded such that the neutral is bonded to both the service enclosure and the earth. Therefore, the residence and the electric

Table 1.1 Ampacity Table for Residential Service

COPPER	ALUMINUM AND COPPER-CLAD ALUMINUM	SERVICE RATING IN AMPS
AWG	AWG	
4	2	100
3	1	110
2	1/0	125
1	2/0	150
1/0	3/0	175
2/0	4/0	200
3/0	250 kcmil	225

ALERT!

When the service amperage and residential service equipment exceed or fail to meet the load being applied, a current overage occurs, causing excessive heat and tripping a protective device like a fuse or a circuit breaker. To avoid this risk, accommodate the load being applied.

utility must agree on a reading of 0 volts. The best way to ensure this agreement is to tie the residence and utility together using Earth as common ground (NEC 250.4).

To achieve this goal safely, the electrician should mount the main service equipment **panel** inside or outside the residence where the service conductors enter the building. In front of the panel from floor to ceiling, NEC requires a clearance width of 30 inches and a clearance depth of 36 inches (NEC 110.26). Due to safety concerns, the panel must be accessible at all times.

Hints and Tips

Due to moisture exposure, accessibility challenges, and poor ventilation, a bathroom, kitchen, or closet is never an appropriate location for a main service equipment panel (NEC 230.70). When service conductors enter a residence at an inconvenient location, consider using a master disconnect and subpanel in a four-wire system per NEC Article 230. To guarantee the panel is always accessible, allow no plumbing, fixtures, or other elements (e.g., cabinets) to reside in front of the panel.

Exploring Branch Circuits

Once it passes through the main service equipment panel, energy feeds several **branch circuits**, which serve specific functions in a residence, including running small appliances. The NEC has requirements for these branch circuits, both in definition and methodology.

Identifying Required Branch Circuit Types

- *Small Appliance.* In a typical residence, the NEC requires at least two 20-amp branch circuits to

Hints and Tips

A service equipment panel has a main panel and **subpanels**. The main portion of the service equipment panel bonds the neutral and equipment grounding conductors. Subpanels separate neutral and ground.

feed receptacles intended for small appliances. Such circuits should be dedicated to areas with small appliances, places like the kitchen (which usually has two) the pantry, and the dining room. Lighting outlets and built-in appliances are not allowed on small appliance branch circuits (NEC 210.11).

- *Laundry.* According to the NEC, one 20-amp branch circuit shall be dedicated for service to the laundry room. Nothing else should be on this circuit (NEC 210.11).
- *Bathroom.* The NEC mandates one dedicated 20-amp circuit for a residence's bathroom **receptacles.** In nearly all geographic areas, dedicated circuits are required for residences with single bathrooms (NEC 210.11).
- *HVAC System.* Following NEC requirements, a heating, ventilating, and air conditioning (HVAC) system belongs on a separate branch circuit in a residence (NEC 424.3). A heating system determines a circuit's ampacity when there is no air conditioning. When air conditioning is present, the larger of the two systems (in terms of ampacity requirements) is used for calculations. The NEC is standardized to use heating load to determine ampacity because more homes have heating units than air-conditioning ones, among other reasons.
- *General.* General branch circuits, which cover residential areas not defined by the preceding branch circuits, should allow 3 volt-amperes per square foot of livable space, according to the NEC. For example, there should be one 15-amp circuit per 500 square feet, or one 20-amp circuit per 800 square feet, to allow for derating and future expansion (NEC 220.12).

Hints and Tips
For more information on HVAC systems, see NEC Article 440.

Studying Branch Circuit Wiring Methodology

The most important aspect of wiring branch circuits is ensuring that the ampacities of the wire, breaker, devices, and load all agree. The conductor of choice for residences, Type NM-B cable, often called Romex, comes in several gauges. Wire gauges range from 14 AWG at the smallest to 0000, or "four naught," at the largest. The latter is used for **service entrances** and large service-to-panel connections.

- *14 AWG 2 conductor with ground for general-purpose 15-amp circuits.* Used mainly in lighting circuits; not robust enough for appliances (NEC Table 310.16). Staple of today's residential wiring installations.
- *14 AWG 3 conductor with ground for special applications*—For three- and four-way switches, as well as smoke detectors. The extra wire is red. Inappropriate for wiring 240V appliances, because it is too fine a gauge to handle the energy.
- *12 AWG 2 conductor with ground for 15- and 20-amp circuits.* Used in 15-amp circuits to allow for future expansion, although expansion would require a breaker change (NEC Table 310.16).
- *10 AWG 3 conductor with ground for 240V appliances, except for the following high-demand appliances:*
    - 8 AWG 3 conductor with ground for ranges and electric ovens
    - 6 AWG 3 conductor with ground for oven appliances

When wiring branch circuits, keep the following guidelines in mind (NEC Articles 210 and 220):

- Circuits rated at particular ampacities must maintain sufficient wire gauge throughout the circuits. An entire circuit must be constructed to accommodate the rating of the breaker used to protect the circuit (NEC Article 210).
- Protection must be commensurate with the ampacity of the wiring, not with the current requirements of the load. In other words, it is not possible to increase a circuit's usable load by replacing a 15-amp breaker with a 20-amp one, unless the conductors are also upgraded to meet the new ampacity (NEC Article 220).
- Affix Romex to a wall or stud using a staple within 12 inches of a junction box, unless the junction box is nonmetallic and single gang. Such boxes require staples at shorter distances, because they offer no structural integrity of their own (NEC 314.17[C] and 334.30).

ALERT!

Staples are a wiring hazard, so use as few as possible during mounting.

- At minimum, staple Romex to a wall or framing at 54 inches intervals. A distance of 3 feet to 4 feet is most safe and effective (NEC 314).
- To prevent rubbing and eventual failure, use proper grommets, clamps, and hardware where Romex enters an enclosure. While the NEC does not specifically state this, failure to do so will void a certification and enclosure listing.
- Mount cabling to framing at least 1.25 inches from the framing's edge (Figure 1.1). When this is impossible, use a wiring guard (Figure 1.2). (NEC Sections 300.4 and 334.15[B])

Figure 1.1 Romex Clearance through Framing

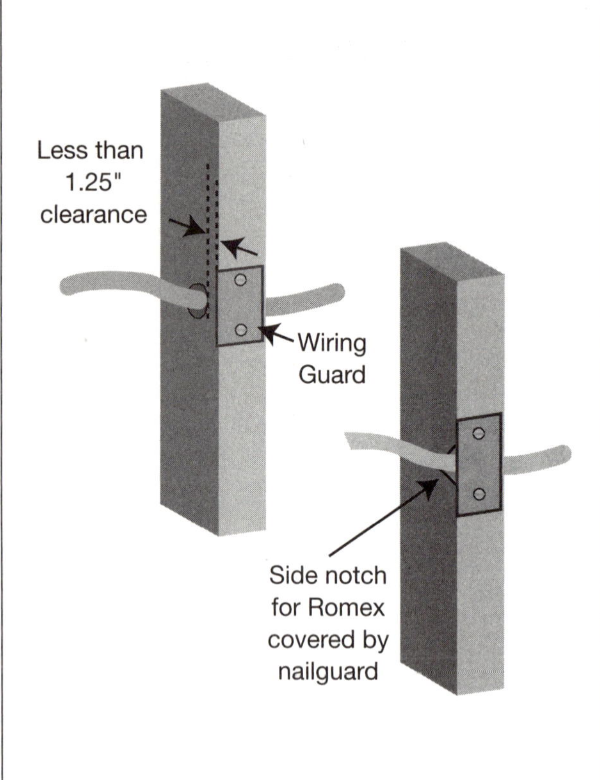

Figure 1.2 Wiring Guard

- Support ceiling-mounted fixtures weighing 35 pounds or less with ceiling-fan-style boxes designed for this use. Many such boxes have cross braces to mount to ceiling joists. Support fixtures weighing more than 35 pounds independent of boxes (NEC 314.27).

ALERT!

Ceiling fans can be particularly hazardous when installed improperly. Their rotational movement causes repetitive oscillation and wear with even the slightest imbalance. Therefore, mount them with a robust method.

Did You Know?

To try to stave off the increasing price of copper in the 1970s, contractors began installing much less expensive aluminum wire in residences. Unfortunately, aluminum has a different coefficient of expansion than copper, which means it behaves differently as it ages and the temperature changes. After several months, aluminum terminations that were secure at installation began to loosen and cause arcing and open circuits. Such phenomena had fatal effects in one home, which was leaking natural gas. When the homeowner returned home and turned on a hallway light, the house exploded, leveling it and killing the owner. Although aluminum is no longer used in home wiring, you may encounter it in a remodeling job. If you do, express your concern and sell the customer a safer alternative.

Ensuring Protection in Residential Installations

A well-designed electrical distribution system has several layers of protection that begin at the utility company and end, sometimes, in the consumer's hands. In general, the closer a protective element is to the consumer, the greater its effectiveness. For instance, circuit protection at the power company will not safeguard a consumer who drops a hair dryer into a bathtub while bathing, but protection on the hair dryer's outlet likely will.

In a residence, the first layer of protection resides in the load distribution center, or panel, in the form of a fuse or circuit breaker. From the panel, the service entrance branches out through several circuits. Each of these circuits is designed to function at a certain amperage and in a certain environment. This design is implemented through choices related to conductor gauge, breaker rating, and increased protection.

A fuse or a circuit breaker is designed to monitor each circuit and open that circuit when the current through it exceeds a certain value. **Fuses**, which consist of small, metallic strips that pass current, are replaceable and disposable (Figure 1.3). When current becomes excessive, the metallic strips heat to the point that they melt, opening or "breaking" the circuit. The homeowner must then unscrew the fuse and install a new one of the same value. The disadvantage of fuses is inconvenience. When you are out of fuses, you are out of electricity.

ALERT!

When replacing fuses, be sure to select identical replacements. Failing to do so could cause overheating, fixture and wiring destruction, or fire. The ratings that appear on all fuses and circuit breakers are the guide to choosing suitable replacements. In addition to having the same ratings as their predecessors, circuit breakers must be the same brands as panels, or the listings are voided.

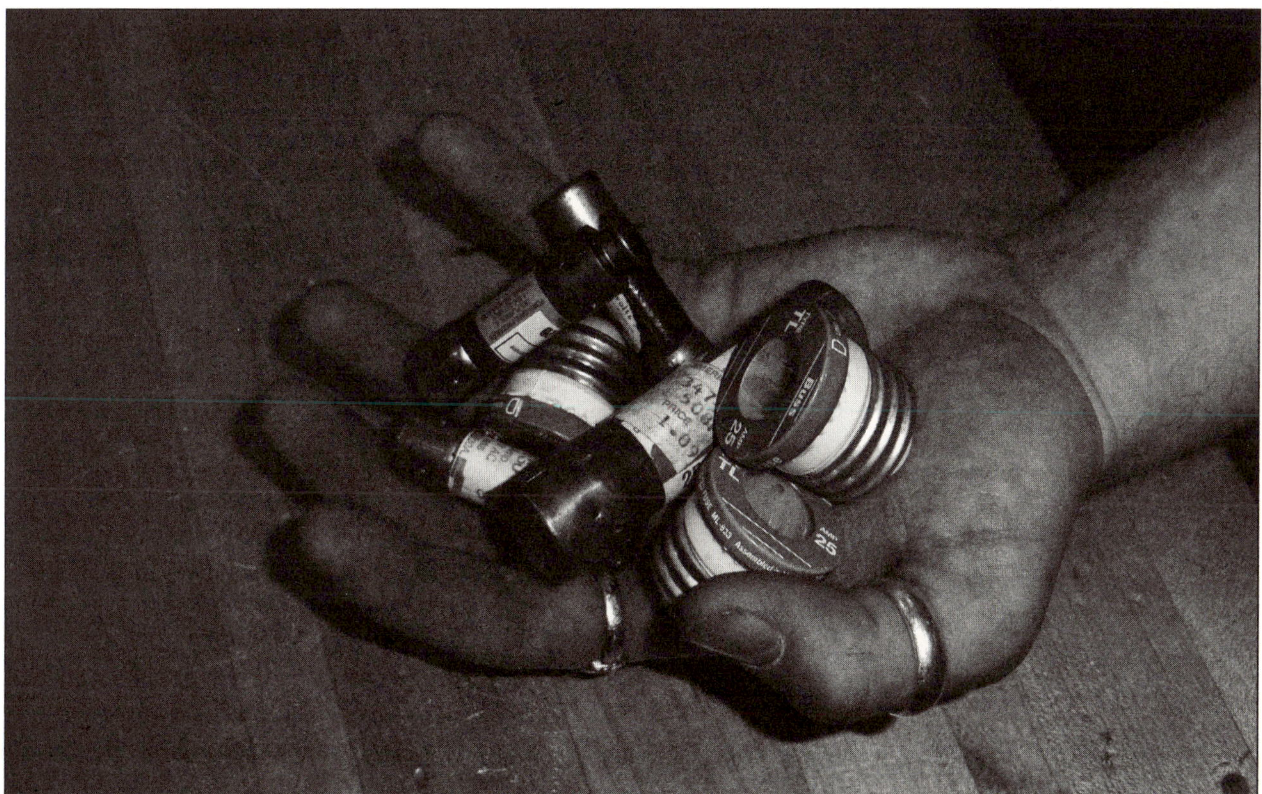

Figure 1.3 Sample Fuses

Figure 1.4 Sample Circuit Breakers

Circuit breakers (Figure 1.4) were developed to address the inconvenience of fuses. Because circuit breakers are resettable, they are ultimately cheaper. They are disadvantageous in that they can occasionally malfunction. Both circuit breakers and fuses are rated in volts and amperes.

Like fuses and circuit breakers, whole-house surge suppression devices offer consumers some level of protection. Because these devices smooth spikes in line voltage, they safeguard electronic devices more effectively than they do people.

Ground fault circuit interrupters (GFCIs) are protective devices that very quickly disconnect a power source in the event of a **ground fault** (Figure 1.5), such as when an extension cord falls into a crowded swimming pool. Specifically, GFCIs detect an imbalance between current on the "hot," or ungrounded, conductor and the neutral, or grounded, conductor. Such

Figure 1.5 Ground Fault Circuit Interrupter

Job Connection

Electrical conductors have not always been coated in plastic, placed in plastic sheaths, and stripped with precision instruments. Many old homes use wiring technology known as "knob and tube." In this approach, white, cylindrical, ceramic insulators, called "knobs" because they resemble cabinetry knobs, are screwed or nailed into structural members. Wires covered in loose-fitting cloth insulation, or "tubes," are fed through and secured to the knobs. The tubes remain an inch or so from the mounting surface.

As time has passed, the knob-and-tube approach has proven hazardous. The tubes, the wires' cloth insulators, become brittle and chip away, posing a risk. As a result, modern breaker panels and Romex have replaced many knob-and-tube installations. However, some do-it-yourselfers and unethical tradespeople replace only visible tubes and leave tubes in hard-to-access ceilings and floors. Jorge Gonzalez specializes in finding and fixing such problems. He has invested in specialized equipment for fishing new wires through walls: infrared imaging, fiber-optic remote cameras, radio frequency (RF) tracking technology, and 20-foot flexible drill bits. He now holds the contract for electrical renovations in all the historical buildings in his hometown, and maintaining those buildings has become his passion.

devices trip when this imbalance exceeds 4–6 mA. Think of a ground fault as a plumbing leak. A GFCI detects such leaks and then shuts off the main supply.

Because both circuit breakers and fuses rely on heat to disconnect a power source, they need time to react. Through this delay, they allow some devices to use more current when starting without tripping the breaker. Such an approach is less than ideal when personal safety is at stake, however. Relying on electromagnetic energy to sense a current increase, a GFCI can respond almost immediately, disconnecting power in milliseconds rather than seconds.

A GFCI receptacle is required in the following six conditions (NEC 210.8[A] and [B]):

1. Bathroom receptacle outlets
2. Any exterior receptacles, including those in the garage
3. Kitchen countertop receptacles, including islands
4. Receptacles within 6 feet of a wet-bar sink with running water
5. All receptacles in an unfinished basement
6. All receptacles at or below ground level

Manual disconnects allow complete, physical disconnects (power and ground) for equipment servicing or replacing (Figure 1.6). In some installations, manual disconnects are required within a certain distance of the devices they disconnect. Such devices must be in sight of the following:

- electric water heaters
- well pumps
- HVAC equipment
- hot tubs
- tanning beds

Figure 1.6 Manual Disconnect

▶ Working with Box Population and Fill

A junction box should be large enough to accommodate the number and sizes of the conductors it connects. To determine the size for a nonstandard box (standard boxes are discussed in NEC 314.1b), the box's volume is calculated. For example, a 2 × 4 × 2 inch box has a volume of 16 cubic inches (2 × 4 × 2 = 16). Each wire entering a junction box is assigned a volume requirement. Except for possibly a factory fixture pigtail, all wires entering a box should be the same gauge, but this is not a requirement. For example, if a 20-amp circuit powers bathroom receptacles but lights and fans are powered via 15-amp circuits, a #12 and a #14 in a 2-gang box could power the sink receptacle and vanity light switch.

According to NEC Article 314.16, which covers box fill:

One #14 conductor requires 2 cubic inches of box space.
One #12 conductor requires $2\frac{1}{4}$ cubic inches of box space.
One #8 conductor requires 3 cubic inches of box space.
One #6 conductor requires 5 cubic inches of box space.

All grounds count as one conductor of the common size.
A receptacle or switch counts as two conductors of the common size.
A yoke or strap with receptacles or switches counts as two conductors of the size connected to that device.

Using the preceding information as a guide, a box with three switches has six NM-B 12-2 sheaths entering to be connected to the switches. The fill requirements are calculated as follows:

6 sheaths, 2 conductors each = 12 conductors
3 switches, each counts as 2 conductors = 6 conductors
Ground counts as 1 conductor = 1 conductor

Total = 12 + 6 + 1 = 19 conductors

19 × 2.25 = 42.75 cubic inches or an approximately $5\frac{1}{2} \times 2 \times 4$ inch box

NEC 314.16 (A) and (A2) discuss metallic and nonmetallic boxes, respectively; nonmetallic boxes must have their volumes stamped into them, while metallic boxes have no such requirement.

Hints and Tips

NEC specifies the minimum acceptable level of work required to be safe, not the best level of work for convenience and customer satisfaction. Therefore, always exceed NEC requirements—always work "above code"—when specifying and installing boxes. It is easier to expand into a large box while a structure is being erected than it is expand into a box once a residence or job is completed.

▶ Grounding and Bonding Wiring Properly

To provide safe and reliable electrical service, the neutral or "reference" leads of residential wiring must be at or as close to 0 ohms as possible. This is done through several methods, often dependent on the physical environment as well as code requirements. The NEC is fairly specific as to how this is to be accomplished, but within those specifications there is flexibility when configuring the grounding system.

A residential grounding system has two parts. The first part is internal. The ground (bare wire or green wire) system effectively connects all boxes and equipment/appliance chassis to panel ground (equipment grounding conductors). The second part of a residential grounding system is the external connection between the main panel/service entrance and the earth (grounding electrode conductor and grounding electrode). This establishes panel ground. Proper internal grounding is accomplished through best wiring practices, detailed in Chapter 3.

The rest of this chapter covers external grounding and the establishment of a grounding electrode outside the home. One may ask, "Why not simply use the house plumbing system for grounding?" After all, the metal pipes enter the ground and are therefore "grounded." The answer is that you can, but per NEC, there must be a dedicated ground connection that does not rely on other incidental ground connections. As an electrician, you will find residential installations that are grounded using a number of unacceptable methods. To meet code, these other methods must be replaced or supplemented by bona fide grounding rod(s) driven into the earth (Figure 1.7).

Once the outside grounding plane is set, the next step is to connect this network of Earth grounds and the service panel. This is done by bonding a wire to the ground rod network and then connecting the other end of the wire to the breaker box or panel. In general, this wire should be robust. A 6 AWG wire is appropriate for small residential service installations, while a 4 AWG is preferred for those installations exceeding 200

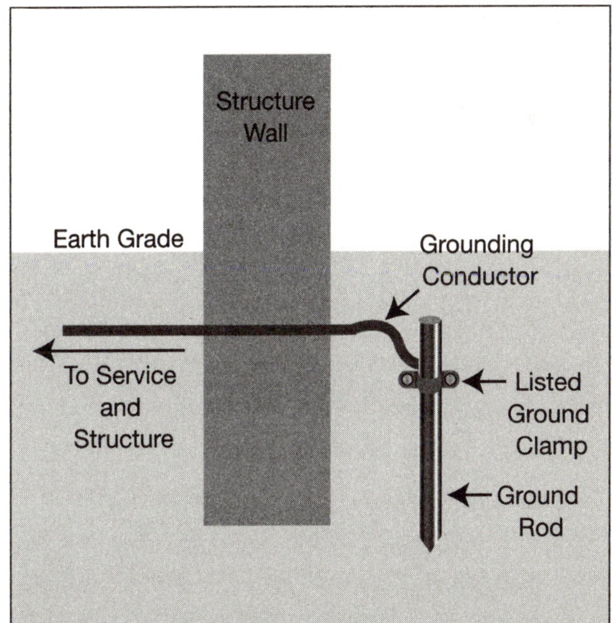

Figure 1.7 Grounding Rod

Hints and Tips
Grounding requirements vary by state and usually depend on the geographical features of the locale. There may, for example, be different requirements in wet, sandy areas as opposed to elevated, rocky service installations. Check with your local agency to learn the specific requirements of your area.

amps of service capacity (NEC Table 250.66). Using direct burial clamps, this wire is clamped or welded to the rod(s) and routed into the residence through the service entrance, where it connects to the fuse or breaker panel (NEC Table 250.64).

The two most common grounding rods are both made of steel; some may be made of copper-clad steel. This grounding rod must be at least $\frac{5}{8}$ inch in diameter, with a minimum length of 8 feet (NEC Section 250.52). In cold areas, the length requirement may be increased to get the rod deep enough to sink below the freezing line (ice conducts electricity less effectively than water). Grounding rods are typically installed by pounding them into the ground at no more than a 45-degree angle and cementing them in place. When possible, the rod's top end should be buried to avoid a tripping or mowing hazard (NEC 250.53[G]).

In some hard-to-ground areas, multiple rods may be used. When they are, they must be separated by at least 6 feet and interconnected using approved underground hardware. The connection to the panel box must also use an approved underground connector to bond the rod to the grounding wire. These systems must be designed and installed in a manner that keeps them reliable for decades through varied environmental cycles and events.

Hints and Tips
Whenever possible while visiting a customer's residence, conduct a panel checkup that requires you to:

1. Look for rust or corrosion on the panel, and check panel clearance and mounting.
2. Check the conditions of fittings and grommets.
3. Remove each circuit breaker, and check the terminals for corrosion, wear, and other issues.
4. Check the box's interior for evidence of rodents or other pests. Advise the customer of the risks associated with chewed wires and nesting material.
5. Check terminations in the box for tightness and proper installation.
6. Clean and dress the panel.

While panel checkups are additional billable services, ultimately, their goal is to keep customers safe. To help cultivate repeat business, always leave installations better than they started.

A grounding system may be established via a number of methods (NEC 250.50 and 250.52). A well casing, for example, is an excellent source of Earth ground, but it is not designed to be electrically stable over time. Concrete-encased rebar, used to erect metal building frames, underground metal tanks, and other local underground systems, may be used to supplement or define grounding systems. The goal is to keep resistance in the grounding system under 25 ohms (NEC 250.56). To achieve or, preferably, exceed that goal, grounding electrodes must be added. Electrodes must be placed below moisture level and separated by at least 6 feet (NEC 250.53). Gas piping and aluminum rods may not serve as grounding electrodes (NEC 250.52 [8][B]).

While all components of an electrical installation are critical, the grounding system carries the bulk of responsibility for safe and reliable residential electrical function. This, combined with the relatively difficult task of reworking the grounding system, makes it imperative that electricians install grounding systems properly the first time.

Did You Know?

The efficiency of an electrode system increases as the distance separating electrodes increases (NEC 250.56 FPN).

▶Chapter Review

1. A home has dimensions of 28 × 52 ft., which includes a 28 × 12 ft. garage. What is the home's total living area in square feet?
   - **a.** 1,120
   - **b.** 1,456
   - **c.** 2,800
   - **d.** cannot be determined

2. How many watts should be allocated per square foot in a residence?
   - **a.** 1
   - **b.** 3
   - **c.** 5
   - **d.** 10

3. A finished basement should have GFCI protection.
   - **a.** true
   - **b.** false

4. A receptacle or a switch with 14-2 entering it counts as ______________ cubic inches.
   - **a.** 2
   - **b.** 4
   - **c.** 8
   - **d.** 10

5. Which of the following is considered a main service ground?
   - **a.** water pipe
   - **b.** ground rod(s)
   - **c.** ceramic well housing
   - **d.** chain link fence post

6. According to good engineering practice, a sump pump requires a dedicated circuit.
   - **a.** true
   - **b.** false

7. How many cubic inches does each wire take for:

   14 gauge ______________________________

   12 gauge ______________________________

   10 gauge ______________________________

8. What factor determines circuit size?

   ____

   ____

   ____

   ____

9. What is an outlet?

   ____

   ____

   ____

   ____

10. What are NEC requirements for a receptacle on a peninsula?

    ____

    ____

    ____

    ____

11. What is the approach for determining the number of receptacles on a 20-amp circuit?

____

____

____

____

12. A whirlpool tub requires GFCI protection.

a. true

b. false

13. What three areas of a residence require 20-amp circuits?

____

____

____

____

14. A 10 AWG wire carries how many amps? (What size breaker?)

____

____

____

____

15. In front of a panel from floor to ceiling, how much clearance depth is needed?

a. 30 in.

b. 36 in.

c. 48 in.

d. 72 in.

16. Is a receptacle needed close to an air-conditioning unit? If so, how close must it be?

____

____

____

____

17. An attic with a furnace requires a receptacle.

a. true

b. false

18. A ____________ -amp breaker goes on a 12 AWG wire.

a. 10

b. 15

c. 20

d. 22

19. Along a wall space, a first outlet should be no more than ____________ feet from a door.

a. 3

b. 4

c. 6

d. 10

20. A metal fence should be attached to a ground rod.

a. true

b. false

21. It is proper procedure to install an exhaust fan and light in a shower area.

a. true

b. false

22. Name some tools for installing ground rods.

____

____

____

____

23. A receptacle should be how high off the floor?

____

____

24. A switch should be ____________ high off the floor.

25. A lighting circuit in more than one room is acceptable electrical practice.

a. true
b. false

26. All new wiring receptacles must be grounded to a neutral bar in the panel box.

a. true
b. false

27. U F wire is normally used for inside

a. house.
b. the ground.
c. garage.
d. overhead service.

28. What is meant by the phrase *shall be*?

____

____

____

____

29. Identify four parties who should be satisfied with the electrical work in a new residence.

____

____

____

____

30. Why is M/C cable required in a cupboard instead of Romex?

____

____

____

____

CHAPTER

2

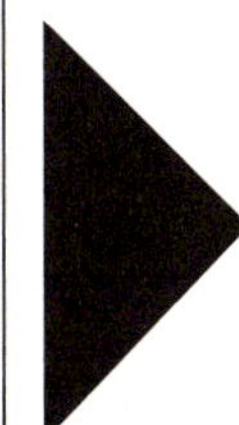

In the Home: Room by Room

CHAPTER SUMMARY

Whether a residence has one story or two, a basement or none, NEC provides guidance for proper electrical wiring. In many cases, local codes provide even more stringent requirements. Living rooms and bedrooms areas are the easiest wiring projects on the continuum, while kitchens pose the challenge of running water. Hallways are unique in that they are purely utilitarian; utility rooms, though underrated, should command due attention. The outdoors, similarly, should not be forgotten. Whatever the wiring task, whatever the area, safety should be the residential electrician's governing concern. Cost and convenience should together follow second.

▶ Learning Objectives

- Discuss the basic principles of wiring a residence
- Name the main elements of a home layout
- Outline the challenges of wiring a bedroom
- Describe how to run wiring in hallways and stairways
- Identify ways to run living room wiring properly
- Talk about ways to wire kitchens safely
- Differentiate wiring practices for utility rooms
- Describe proper wiring practices for bathrooms
- List wiring practices for basements and garages
- Walk through procedures for wiring the outside of a home

▶ Key Terms

appliance
cove lighting
hue
kitchen
light temperature
living room
redundancy
service area
three-watt rule
utility room

▶ Understanding General House Layout

People choose to live in many different types of dwellings. Most of those dwellings fall within a general residential category that includes single- and two-story homes with and without basements. Though this chapter targets those types of dwellings, the National Electrical Code (NEC) outlines best practices for wiring all types of structures, including those being converted to residences.

> **Hints and Tips**
> When commencing any type of wiring work, remember to consult local codes as well as the NEC. In some cases, local guidelines are more stringent than national ones.

Any wiring plan should start with a review of home layout. Most home floor plans identify living, sleeping, and service areas. Living and sleeping areas are generally easiest to lay out from a wiring perspective, because their required level of protection is less stringent than that for **service areas**, places like kitchens, bathrooms, and basements. Places like kitchens and bathrooms require wiring that accommodates machinery and running water. When wiring any room in the home, safety is the primary concern, followed by convenience and then cost.

The lights and receptacles in living areas are typically sourced by one or more general-purpose branch circuits. As discussed in Chapter 1, general-purpose circuits typically have 15- or 20-ampere capacity, and typical home installations are budgeted using the **three-watt** rule, that is, three watts per square foot for lighting circuits.

> **Hints and Tips**
> To deliver power consistently in living areas, distribute circuits so that a room has alternate sources of power. If, for example, a television trips a circuit breaker as it falls onto the floor in a room with distributed circuits, another source of light is available to navigate the shards of broken glass.

▶ Laying Out Bedroom Wiring

In wiring terms, living areas like bedrooms are the simplest projects to lay out because they generally lack such challenges as running water. Architects who are designing new homes speculate on the use of various rooms, identifying where articles of furniture will reside and laying out electrical systems accordingly. A bed, for example, may rest along a room's longest wall, and outlets may appear on either side of that bed where nightstands traditionally go.

In general, at the entrance to a bedroom there should be two switches: (1) one that controls an overhead, utilitarian light fixture and (2) another that controls a switched receptacle. A dimmer on the latter can vary the amount of lighting that is delivered to achieve softer, mood lighting. Both switches should be placed on separate circuits to ensure **redundancy**.

In addition to two switches at a bedroom's entrance, there should be an outlet near each corner and at 12-foot intervals around the room so that no point along the floor line is more than 6 feet from a receptacle (per NEC 210.52[A][1–2]) (Figure 2.1). Outlets near windows allow for holiday lighting. Optional,

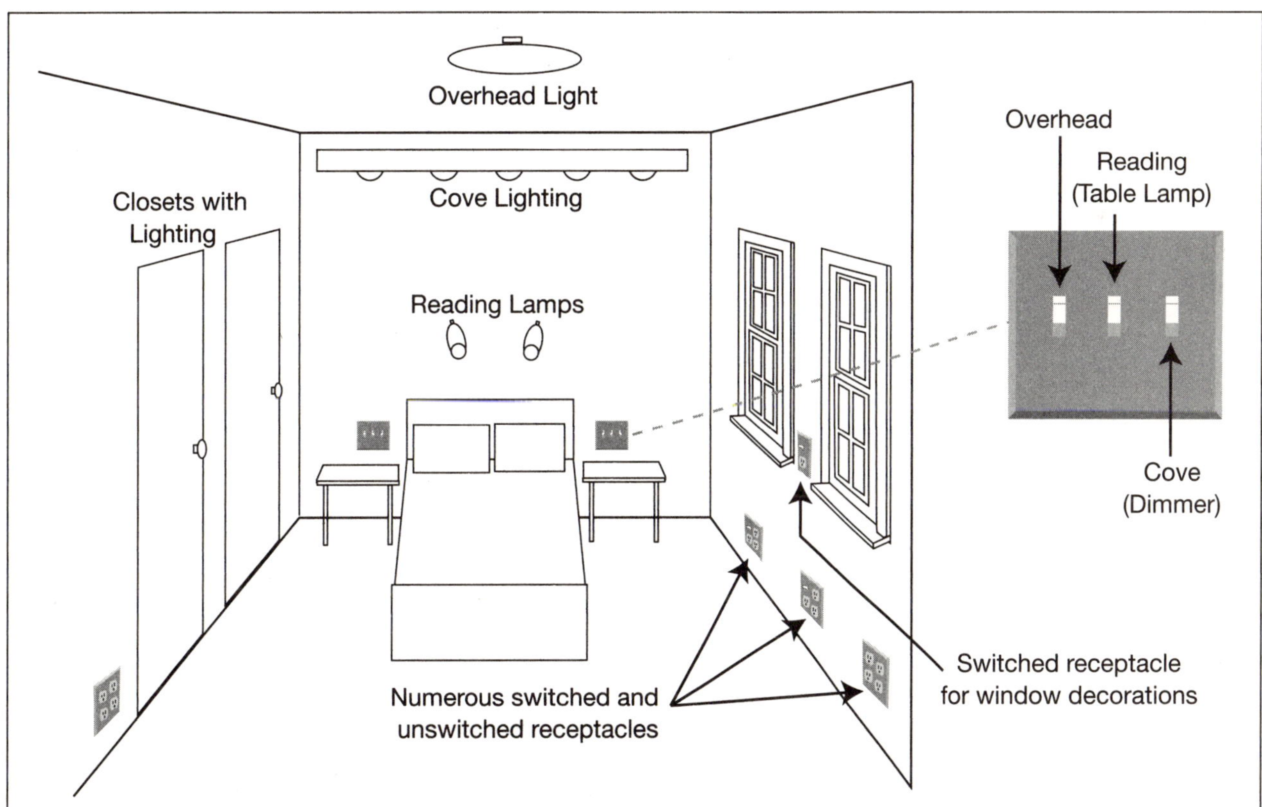

Figure 2.1 Properly Laid Out Bedroom

Job Connection

As you progress as an electrician, you will develop certain shortcuts to become more efficient. When Jules was wiring one bedroom and Samantha was wiring a second, Jules noticed that Samantha always seemed to finish before he did. Seeking clues, he watched Samantha install her receptacle boxes. Whereas Jules carefully measured each box into position, marking the stud with a tape measure and then carefully aligning the box to that mark, Samantha used a piece of wood cut to proper length for reference. She would simply lean the wood piece against the stud, one end on the floor, and then place the box on the other end and hammer it in. Samantha measured once but used it again and again. Because she did, she could wire two bedrooms in the time it took for Jules to complete just one.

Hints and Tips
In bedrooms, provide more receptacles in more places to allow homeowners to configure their rooms as desired.

three-way switching can allow occupants to control lighting from bed. Switched lighting should be installed in bedroom closets. Given their sizes, walk-in closets should be well illuminated.

▶ Running Wire in Hallways and Stairways

Hallways are unique residential areas in that they are purely utilitarian: They simply facilitate transit between two points. Hallways should have three- or even four-way switching so that any room emptying into the hallway can control hallway lighting. The NEC requires a receptacle in a hallway that is 10 feet or longer (Figure 2.2) (NEC 210.52[H]). Stairway requirements include that any staircase with more than five steps have three-way switching at top and bottom (NEC 210.70[A]) and that any landing with a door have a switch for lighting (NEC 210.70[A][1]). Receptacles for decorating stair rails or powering lamps or vacuums at landings are optional.

Did You Know?

The **temperature** of an area's lighting, or **hue**, sets the area's mood. Hue is measured in degrees Kelvin. Blue fluorescent lighting is called "cool" lighting; incandescent lighting is considered "warm."

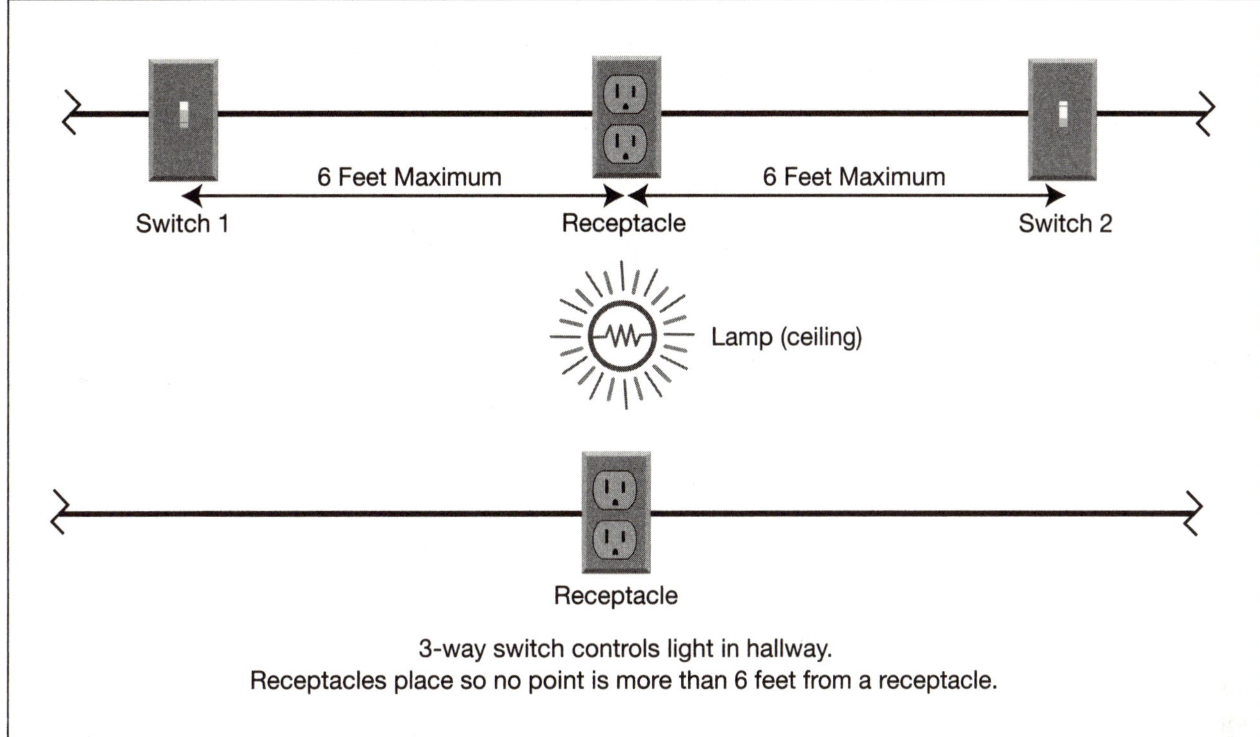

Figure 2.2 NEC-Compliant Hallway

▶ Wiring the Living Room

Per NEC, the term **living room** applies to living rooms proper as well as to recreation rooms, "great" rooms, dens, studies, and areas used for similar purposes. Living rooms require three- or four-way switched lighting at each entrance (NEC 210.70[A][1]), because oftentimes they open to other areas, including dining rooms, bedrooms, and the outdoors.

Like bedrooms, living rooms should have two lighting systems: overhead or utility lighting for vacuuming and other activities; and low-level or dimmable lighting for relaxing. Also, like bedrooms, living rooms should have plentiful receptacles spaced at usable intervals, more than NEC's 6-foot requirement suggests (NEC 210.50[A][1-2]). To power things like holiday lighting, standing lamps, and vacuums, receptacles should reside on either side of windows and in corners. Abundant receptacles, more than those required by code, help prevent improper extension cord use.

> **Hints and Tips**
> When completing all electrical tasks, like installing receptacles, strive to work "above code" by exceeding NEC standards. Doing so enhances both safety and customer service.

Entertainment-center receptacles should be placed on dedicated 20-amp circuits to provide low-voltage wiring for Internet, cable, and computer access (Figure 2.3).

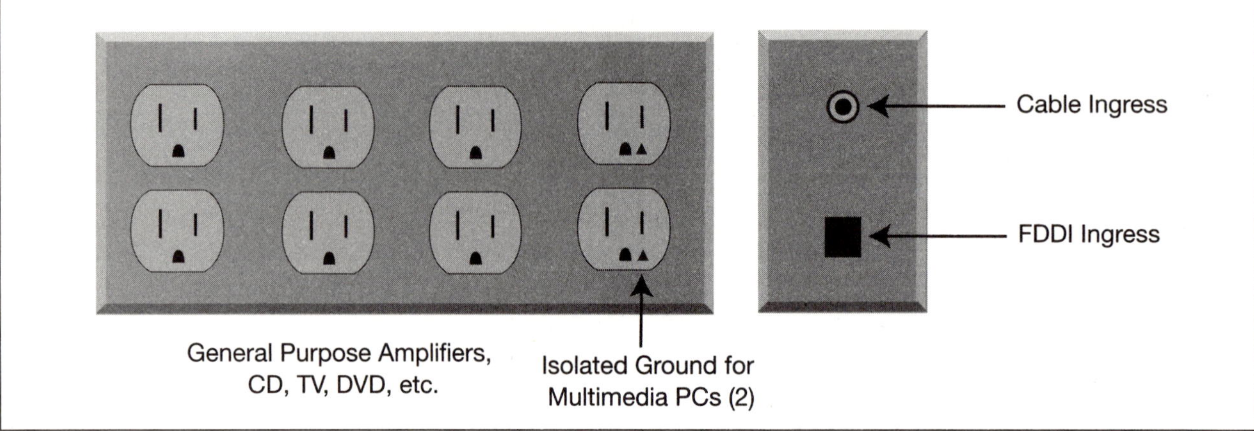

Figure 2.3 Entertainment-Center Wiring

▶ Running Kitchen Wiring

The NEC definition for the term **kitchen** includes areas like the pantry, dining room, and breakfast nook, as well as the kitchen proper. Because kitchens typically house a number of high-demand applications, proper wiring is more important in the kitchen than in any other room in a house. To ensure safety and reliability, a kitchen's electrical system should be well thought out. How much power will the stovetop and oven consume? Will the refrigerator be able to run consistently on 100% duty cycle? What are the power requirements of plug-in **appliances**?

Lighting is a particular concern in kitchen applications, because it helps facilitate safe food preparation. It falls into four basic categories (Figure 2.4):

1. *General*—Usually fluorescent, because incandescent light at the volume required would generate a large surplus of heat energy as well as an excess of power. Most kitchens feature large single fixtures or several ganged fixtures mounted in the centers of their ceilings. This approach leaves many dark areas at the peripheries.
2. *Track*—Useful for lighting countertops in front of the food preparer, who is usually blocking ceiling-mounted general lighting. Low-voltage halogen bulbs are popular with track lighting, because they target light where needed. Newer light-emitting diode (LED) technology is also gaining prominence in this area. In some cases, recessed lighting is used as an alternative to track lighting, as is pendant lighting.
3. *Undercabinet*—Illuminates areas under kitchen cabinets where food and appliances are stored and labor is done. Small halogen fixtures and fluorescent mini-tubes are popular alternatives.
4. *Cove*—Directs light upward. **Cove lighting** is generally installed on top of kitchen cabinets to add character but lacks true function.

Whatever its type, kitchen lighting is generally sourced from one 15- or 20-amp circuit. General or overhead lighting should operate from each kitchen entrance using three- and four-way switching. Auxiliary aesthetic lighting (i.e., undercabinet and cove types) should be controlled from wherever convenient. Track lighting switches should be near the elements they are designed to illuminate.

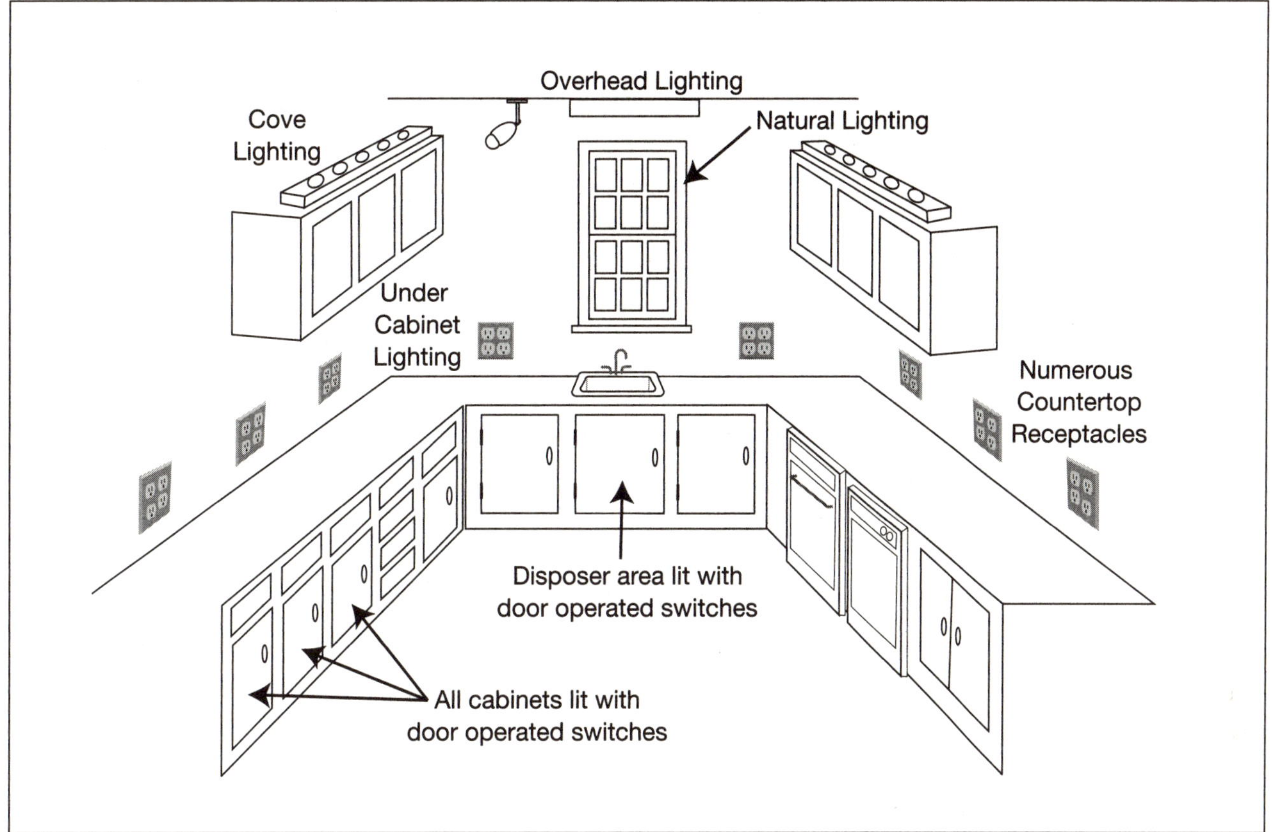

Figure 2.4 Types of Modern Kitchen Lighting

Given their importance in the home, both NEC and local codes have many requirements for kitchen lighting and receptacles. Two separate 20-amp circuits are used for wall, floor, and countertop outlets (NEC 210.52[B][1]), because kitchen loads can be quite high and require a great deal of current. These branch circuits must be dedicated to the kitchen area (NEC 210.52[B][2]), although in some jurisdictions extra switched receptacles belonging to a general-purpose branch circuit may be installed in these areas (NEC 210.52[3][B]). Ground fault circuit interrupter (GFCI) receptacles are required in areas prone to moisture (NEC 210.8[4] through [8]). Kitchen wall outlets should be spaced so that no point on a wall is more than 6 feet laterally from an outlet (NEC 210.50[A]). Per NEC 210.52(B)(3), no small-appliance circuit may source more than one kitchen. NEC 210.11(C)(1) outlines the dedicated branch circuit requirement for kitchen areas.

To decrease the need for extension cords in kitchens, countertops must have receptacles at close intervals (NEC 210.50[C][1]). Many portable kitchen appliances have very short electrical cords to reduce exposure to risk and hazards. Above a kitchen countertop,

> **Hints and Tips**
> When a kitchen has a sliding glass door, for outlet-allocation purposes, the door's fixed portion is considered wall space.

Hints and Tips
When placing receptacles above and around kitchen countertops, be sure to check local code requirements, because those requirements tend to vary between jurisdictions.

NEC requires receptacles to be spaced every 4 feet and no higher than 20 inches above the counter surface (NEC 210.52[C][1] and [C][5]) (Figure 2.5).

When a kitchen has an island that exceeds a surface area of 2 square feet (2 × 1 foot or larger), that island must have receptacles on its surface (NEC 210.52[C][2]). Typically as a best practice, island receptacles are mounted no more than 12 inches below islands' surfaces to minimize the effects of spillages.

For the most part, most large kitchen **appliances** should run on their own dedicated circuits (NEC 210.50). A refrigerator requires its own circuit, because otherwise another appliance could trip the refrigerator's circuit and cause food spoilage. A stove and/or range must occupy its own 240V circuit; for a gas stove, a 120V circuit is appropriate. Dishwashers and garbage disposals are often placed on the same circuits, because they are rarely used at the same time. While dishwashers need not be switched, garbage disposals are always switched (not by code, but because of their configuration), usually from a countertop near a sink.

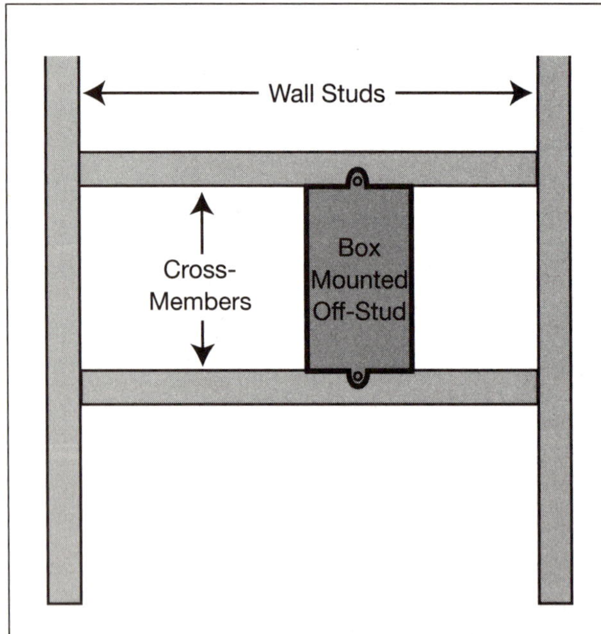

Figure 2.5 Cross-Members that Increase Kitchen-Countertop Receptacle Count and Placement Options

Did You Know?

Fine print notes (FPNs) are supplemental materials that are not regulatory but that offer insight into how requirements are to be interpreted. For example, an FPN allows a refrigerator to power off its own circuit, but it is not required (NEC 422.16).

Ampacities for kitchen appliances are typically:

- Electric stove: 6 AWG copper wire, 50-amp breaker
- Dishwasher: 12 AWG dedicated circuit, 20-amp breaker
- Garbage disposal (often paired with dishwasher): 12 AWG, 20-amp breaker

Hints and Tips
Homeowners today often use more high-wattage appliances than NEC accommodates. Therefore, instead of placing several receptacles on one GFCI circuit, allocate several GFCI circuits, each supplying fewer receptacles.

Hints and Tips

Some local building codes amend the NEC when it comes to utility areas, especially laundry rooms. Therefore, always consult the local building code before wiring a laundry room. When laying out service, try to reduce the need for extension cords by providing receptacles in commonly needed places.

- Refrigerator: 14 AWG dedicated circuit, 10-amp breaker
- Lighting: must be switched and not on one of the small appliance circuits
- Gas stove: can be tied into a general-purpose circuit or on one of the small appliance branch circuits; electricity requirements are low and generally include only the lighting and electronic control system (e.g., clock, timer)

▶ Putting Utility Rooms Together

Utility rooms, generally one of the most significant yet underrated areas in today's homes, include laundry rooms, workshops, and attached garages (NEC 210.70[3]). Many times, utility areas remain unfinished: Floors are often bare concrete, drywall may not be installed or painted, and wiring may be basic (one central light fixture may be supplied). In general, a utility room is wired according to its intended use.

Executing Wiring Plans in Laundry Rooms

A laundry room, one type of utility room, typically needs power for a washing machine, a dryer, and possibly an iron (Figure 2.6). NEC mandates a 120V/20-amp circuit for the laundry, and the laundry must have no other outlets. When homeowners desire more outlets

Figure 2.6 Typical Laundry-Room Appliance Cluster

in the laundry area, those outlets must originate from another circuit (NEC 210.52[B] and 210.9[C][2]).

Although NEC does not require it, receptacles for washing machines and electric clothes dryers should be above and behind the appliances to maximize safety. Use GFCI receptacles with them, and ensure that the washer's water discharge and powering outlet are isolated from one another. Wire an electric clothes dryer on a dedicated 120/240V receptacle. Because gas dryers require minimal electricity relative to electric models, they can be attached to the same general-purpose electrical circuit in the laundry room (NEC 210.11[C][2]). All electrical receptacles used for laundry equipment in a laundry room should be wired on a dedicated 20-amp circuit, but only one receptacle is required (NEC 210.11[C][2]). Consider providing additional receptacles for the homeowner's convenience.

A ceiling-mounted, 4-foot fluorescent fixture should provide sufficient illumination for a typical laundry room. This fixture can share a circuit with an adjoining room. Laundry-room lighting can be on a general lighting circuit; it cannot be on the dedicated laundry-room appliance loop (NEC 210[C][3]).

As mentioned earlier, many laundry areas are found in unfinished basements. When this is the case, a separate GFCI appliance receptacle must be installed in addition to the GFCI receptacle required in the basement.

ALERT!

In a workshop, electrical cords are not only electrical hazards, they can be choking or cutting hazards, as well.

Did You Know?

Many workshop power tools have motors that draw more current at startup than they do while running. Consider this feature when specifying the protection for circuits driving these tools.

Running Wires in Home Workshops

Like a laundry room, a home workshop is a type of utility room. In many ways, workshops are wired like kitchens. Workshops have high-powered appliances and require extra lighting. Workshop receptacles should be plentiful and placed in close proximity along walls both to keep cords short, as most shop appliance cords are, and to avoid extension cord use.

As in a kitchen, proper illumination of workshop work surfaces is critical. It is difficult for an operator to cut a line with a miter saw accurately when standing between the workpiece and overhead lighting. Undercabinet lights or recessed lighting luminaires, or "cans,"

ALERT!

To avoid the lighting outages that risk operator injury, dismemberment, or death, any workbench should have at least six receptacles, all GFCI protected. Additional receptacles help ensure that workpieces are always sufficiently illuminated, particularly when using cutting or shearing appliances.

can solve this problem. The key to proper and effective workshop lighting lies in eliminating shadows, which entails directing light onto a workpiece from several sources. Low-voltage halogen or LED lighting can accomplish this goal.

Hints and Tips
To provide maximal protection to home occupants, provide more than one GFCI circuit in a bathroom.

▶ Meeting Electrical Needs in the Bathroom

In terms of wiring needs, bathrooms and kitchens are similar in one respect: Both have running water and therefore must be safe for human use. NEC requirements for bathrooms include one 20-amp circuit that is GFCI protected at the breaker or receptacle, to which all bathroom electric function is wired via receptacles (NEC 210.11[C][3]). When just one bathroom exists, it is permissible to use the single branch circuit to supply other equipment in that bathroom (NEC 210.11[C][3]). Three bathrooms require three circuits. Alternatively, basin receptacles (GFCI) can be placed on dedicated circuits and remaining bathroom receptacles may be on general purpose circuits.

For heating options in this area of the home, consider combination infrared bulb/vent assemblies or forced-hot-air heaters on the bathroom walls. Usually, such heaters are installed and powered by a dedicated 240V circuit.

Job Connection

In Norfolk, VA, Charley Carter was running wires in the attic of her customer's detached garage one crisp autumn day, when she noticed that part of the garage seemed warmer than the rest. She wondered where this extra heat was coming from and quickly checked to see if she had wired something incorrectly or had inadvertently overlooked a heater or an appliance.

Through her investigation, Charley identified a small cardboard box in one corner that seemed to be the source of the heat she had noticed. Having been in the home a few hours earlier, she remembered that the homeowner had recently applied linseed oil to his random-width hardwood flooring. The half-empty bottle of linseed oil in the attic next to the box was her clue. In the box were all the rags the homeowner had used to apply the linseed oil.

Recognizing the safety hazard, Charley immediately removed the now-near-combustion box of rags and discussed the hazard with the homeowner. In so doing, she addressed a situation that, left unattended, would have likely escalated into a fire in the garage. All electricians and tradespeople should serve their customers likewise.

Other wiring requirements and/or common practices for bathrooms include

- lighting controls by the bathroom door (NEC 210.70 [A][1]).
- receptacles within 3 feet of the sink either on the wall behind the sink or on the front of the vanity not more than 1 foot below the top surface. Double sinks can share a receptacle when both sinks are within 3 feet of the receptacle (NEC 210.50[D]).
- Vent fan when the bathroom lacks an openable window, which is a local requirement in most areas. Vent fans in shower or bath areas must be listed for wet use and GFCI protected (NEC 410.90[D]).
- Lighting fixtures no less than 3 feet from a tub, except when the bathroom has an extremely high ceiling that keeps the fixtures 8 feet above the tub (NEC 410.10[D]).

ALERT!

When a garage is attached to a residence, the joining or common wall is usually fireproof, a "firewall" that relies on its integrity to function. Given that, do *not* install a receptacle or panel on this wall directly across from a receptacle or panel in the home's interior, because doing so provides a route for fire to travel through the wall and into the house.

▶ Installing Wire in Basements and Garages

The wiring requirements for unfinished basements and attached garages are straightforward: at least one GFCI-protected receptacle and one switched lighting fixture (NEC 210.50[1] and [2]).

▶ Placing Wiring Outside the House

For wiring purposes, areas outside the home include crawl spaces, which are under the home, and attics, which are above the home.

According to the NEC, homeowners must have two external receptacles: one in front of the residence and one in the rear (NEC 210.50[E]). Such receptacles can be flush mounted or recessed (Figure 2.7), but both must be GFCI protected and certified for outside

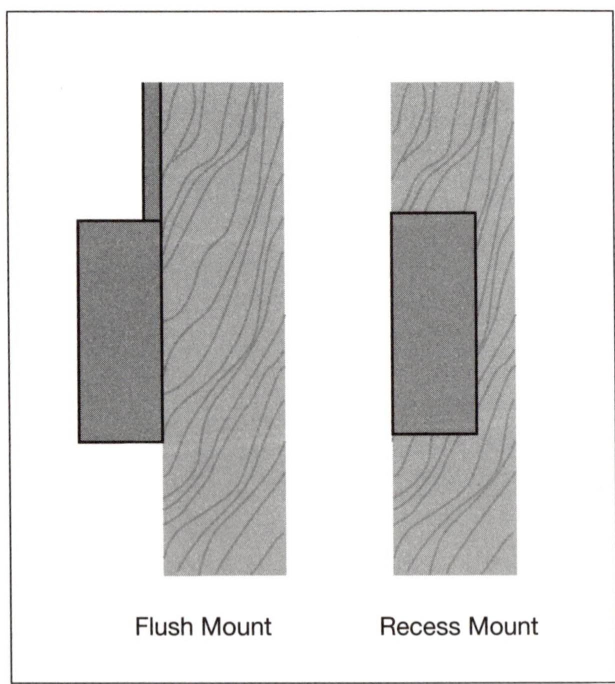

Figure 2.7 Flush- and Recess-Mounted External Boxes

use, meaning they should be waterproof (Figure 2.8) and within 25 feet of an entrance.

Decking on a residence renders it acceptable to increase the number of receptacles outside. While outdoor receptacles should operate on a dedicated circuit (NEC 210.50[E][3]), they can share the circuit used for exterior lighting. Exterior lighting is required only for entry points to a home, places like front and back doors. These lights need not be switched, but most homeowners prefer to be able to turn their exterior lights on and off.

The presence of serviceable equipment determines the code requirements for attics and crawl spaces. When serviceable equipment is present, a switched light must exist at the point of entrance. For an attic with a fold-down staircase, for example, lighting can be switched from the wall on the floor below the attic or from the inside entrance to the attic. Wiring within 6 feet of an entrance must be physically protected against damage (NEC 320.23[A]).

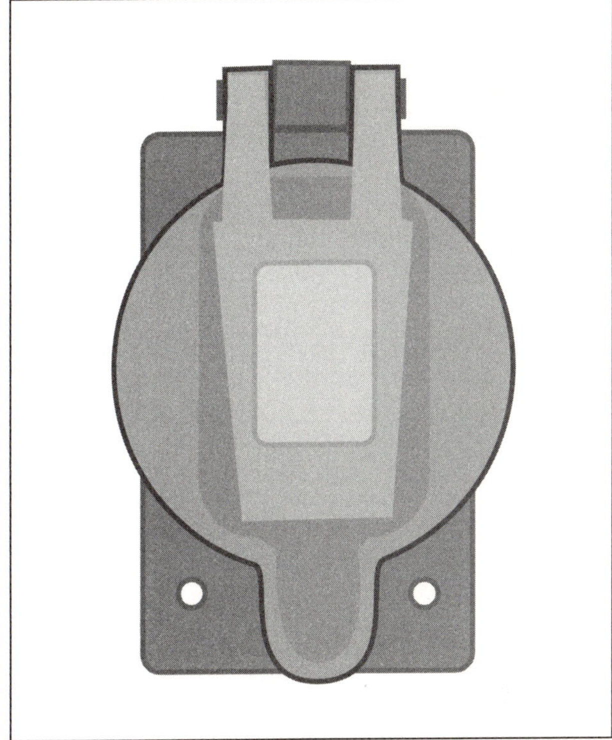

Figure 2.8 Waterproof Receptacle Cover

▶Chapter Review

1. A kitchen island requires a receptacle when
   a. any side of the island is longer than 3 ft.
   b. the island's surface area is 2 sq. ft. or greater.
   c. a sink is present.
   d. No requirement governs island power.

2. List two common types of kitchen lighting.

   ____

   ____

3. An electrical service requires ________ ground rods that are ________ long and ________ far apart.

4. When you install a receptacle indoors near a patio door, the movable section receptacle should be within 8 ft. of the door.
   a. true
   b. false

5. A light switch is needed at each doorway entrance.
   a. true
   b. false

6. A wall space ____________ long must have a receptacle.

7. On what size wire does the dining room receptacle circuit need to be?

   ____

   ____

8. It is good practice to place a light above a kitchen sink.
   a. true
   b. false

9. A sump pump should run on a dedicated circuit.
   a. true
   b. false

10. An electric stove takes ________ size wire and ________ size breaker.

11. An 18-ft. wall with a door at either end needs how many receptacles?

    ____

    ____

12. A 3-ft. wall directly inside a bedroom requires a receptacle.
    a. true
    b. false

13. A kitchen countertop needs how many circuits?

    ____

    ____

14. Kitchen counter circuits require what size wire?

    ____

    ____

15. The receptacles on a kitchen counter should be not more than ____________ apart.
a. 6 in.
b. 8 in.
c. 1 ft.
d. 4 ft.

16. A garbage disposal requires a cord.
a. true
b. false

17. What is the smallest counter size that needs a receptacle?

____

____

18. What size circuit is the GFCI receptacle in a bathroom?

____

____

19. A window candle switch outlet in a side bathroom requires GFCI protection.
a. true
b. false

20. Give an NEC reference for an unfinished basement.

____

____

21. Identify six areas in a home that require overhead lighting.

____

____

____

____

____

____

22. A refrigerator should run on its own circuit in the kitchen area.
a. true
b. false

23. How many receptacles are needed outside a home and where?

____

____

____

____

24. In a closet that is 3 ft. deep and 6 ft. wide and that has a shelf 20 in. wide, what type of light is appropriate?
a. incandescent
b. track
c. cove
d. fluorescent

25. A 9-ft. countertop requires how many receptacles?

a. 1
b. 2
c. 4
d. 8

26. It is proper practice to install a duplex outlet for a refrigerator.

a. true
b. false

27. In a bathroom with a window, an exhaust fan is needed.

a. true
b. false

28. A kitchen counter circuit is usable for a gas stove.

a. true
b. false

29. What size wire does a clothes dryer need? How far from the dryer can the plug be?

____

____

____

____

30. A 20-amp circuit powering a washer can have more than one duplex receptacle.

a. true
b. false

CHAPTER

3 Residential Wiring Techniques

CHAPTER SUMMARY

Sound residential wiring is founded on basics. Once electricity travels from a distribution system to a transformer, it must be properly incorporated into a residence to safely power lights and appliances. Whether delivered overhead or underground, electricity must meet residential service requirements and be properly grounded. Successful electricians chart service road maps at the start of installations to guide the process of choosing correct panels, creating the service entrance, connecting the meter base, and establishing the wiring infrastructure that properly harnesses the panel, conductors, junction boxes, and switches, as well as any outside fixtures and elements. With an installation's infrastructure in place, the electrician can set about executing fitting and finishing techniques. Testing and verification render the process complete.

▶ Learning Objectives

- Explain how voltage is supplied to a residence
- Compare and contrast underground and overhead service
- Discuss proper panel specification and installation
- Explain how to run Romex from panel to box
- Explain how to terminate conductors
- Calculate box size based on conductor size and population
- Discuss how to install a grounding system

▶ Key Terms

arc
contact
derate
drip loop
flush mount
ground fault circuit interruption (GFCI)
junction box
line voltage
mast
megger
meter base
neutral
new work
old work
overhead service
panel
pigtail
pole
punchout
recessed mount
service entrance (SE) cable
service head
service point
splice
strap
switch
traveler
underground service
wire nut

This chapter tackles residential wiring techniques starting with the basics. First is a discussion of how electricity is manufactured, from a distribution system to a transformer and then to a residence. The next section examines how electricity is delivered to lights and appliances. Throughout, proper residential electrical pathways are mapped from point to point. Final sections describe how to connect conductors safely and effectively and how to execute fitting and finishing techniques.

▶ Introducing Electricity Basics

Electricity is manufactured mechanically, through hydroelectric dams and turbines or coal, oil, and other fuel-related turbines, or it is made chemically, through means like batteries and solar arrays. Usually, power travels from an underground or overhead substation through neighborhoods and along streets via overhead means (i.e., **poles**) or underground ones (i.e., filled ditches).

Both underground electrical service and overhead electrical service have unique advantages. **Underground service** requires less tree work during installation and maintenance, has no roofing implications, and is exempt from weather concerns. **Overhead service**, in contrast, is simpler to troubleshoot and repair and is the easier option when running electricity across roadways, waterways (e.g., gulleys, creeks, or streams), or right-of-ways.

Whichever way electrical service is delivered, it leaves the power plant in a high-voltage, low-current state. As it traverses power lines to its destination (e.g., a residence), it passes through transformers that both bring down its voltage and increase its current. By the time it reaches a residence, electricity is at "line voltage," or 120V per phase. **Line voltage** is the voltage in a home's wall outlets. Typically these are seen as two separate phases which, when combined with a multipole breaker, yield 240V for higher-powered appliances.

Three wires enter a home via its service entrance: (1) black (120V), (2) white or bare (neutral), and black with a red stripe (120V). When each of the 120V wires is measured against the neutral one, the voltage is 120V. When one of those 120V wires is measured against the other, the result is 240V (120V and 120V), the voltage needed to power such appliances as dryers and electric stoves.

When establishing electrical services, electricians connect the neutral and the earth. Down the line, they connect the **neutral** to what the power company considers 0 volts. To ensure that neutral represents Earth ground at the residence, electricians must connect the neutral line to the house's ground system, usually via a combination of grounding rods and well and plumbing/sewer pipes, if those pipes are metal.

▶ Determining Service Requirements

Electricians start wiring jobs at a home's meters, or service point, coordinating with power companies to ensure that electrical service is disconnected before beginning work. NEC defines **service points** as the point at which the homeowner takes responsibility for wiring and usage (NEC Article 100).

The first thing an electrician does is establish a home's service requirements, in amperes, using Appendix D of the NEC. Recall from Chapter 2 the 3-watts-per-square-foot requirement for household lighting (NEC Article 230). Using this rule, the service requirement for a 1,000-square-foot house would be calculated as (NEC 220.12):

$$3 \text{ watts/sq. ft.} \times 1{,}000 \text{ sq. ft.} = 3{,}000 \text{ watts}$$

With a home's service requirements identified, the electrician's next task is to examine the home's floor plan and map out branch circuit requirements based on receptacles and appliances. The electrician first maps receptacles to the branch circuit requirements of each room, typically allowing 20 amperes per circuit. The electrician then determines appliance requirements using the appliances' labeled ratings. Following are sample branch circuit requirements (NEC 210.23[A]):

Lighting/general	3,000W
Electric stove	6,000–8,000W (check plate)
Electric water heater	3,000–8,000W
Dryer	5,000W
Kitchen circuits (2)	3,000W
Laundry circuit	1,500W
Disposal	1,000W
HVAC	15,000 VA

Because these requirements are not for concurrent use, the electrician can **derate** them to determine service requirements as follows (NEC Article 220):

1. First, add all wattage required, except for HVAC.

   3,000W + 8,000W + 8,000W + 5,000W
   + 3,000W + 1,500W + 1,000W
   = 39,500W or 40,000W

2. Calculate 25% of the wattage required and 50% of the remaining watts.

   25% × 40,000W = 10,000W

   50% × 30,000W = 15,000W

3. Add the two figures.

   15,000W + 10,000W = 25,000W

4. Add HVAC wattage requirements.

   15,000W + 15,000W = 30,000W

5. Divide by supply voltage to calculate service requirements.

   30,000W ÷ 240V = 125 amperes

To allow for future expansion and/or code requirements, the electrician should specify 200-amp service for this installation, which is standard in contemporary residences.

▶ Selecting the Correct Panel

Done well, the exercise that determines service requirements yields a road map for laying out an electrical service. After determining service requirements, the electrician's next step is to choose an appropriate breaker box, or **panel**, considering two major factors: circuits and manufacturer.

Hints and Tips
Choose panel manufacturers that are reputable and appear to be viable for the long term.

When sizing a panel, the number of circuits is independent of total panel ampacity. For example, two panels may each supply 100A; one may have five 20A 120V circuits, while the other may supply just two 50A circuits. Electricians should be sure to accommodate expansion for extra circuits. Recall that the recommendation for the electrician in our example was to size the service for 200 amps when only 125 amps were required. Panel capacity should increase correspondingly.

In terms of manufacturer, NEC requires the panel and circuit breakers in an installation be made by the same company (NEC 110.3[B]).

▶ Creating the Service Entrance

Once an electrician has chosen an appropriate panel, it is time to create a service entrance, which includes a meter and meter-to-panel wiring as well as the panel. The service's ampacity, which is 200 amps in our example, determines which **meter base** and wiring the electrician should choose.

When establishing a service entrance, the electrician must choose an accessible outside location free of blocking structures like gates or other lockable barriers (NEC 230). To minimize the wire needed, which is expensive and difficult to manipulate, the electrician should insert the inside panel through a wall into the house (NEC 230.4). For most overhead service, electricians install **drip loops** between service poles and res-

Figure 3.1 Simple Drip Loop

Did You Know?

Local guidelines vary as to who is responsible for which portion of service-entrance wiring. In general, meters plug in to standard meter bases. Some power companies provide those meter bases and then install the corresponding wiring from the service pole or underground feed to the meter base. In cases like these, the electrician is responsible for all service from the meter base into the residence. In other cases, the electrician and not the power company is responsible for the meter base. Check local guidelines for specifics.

idences to keep water away from the meter and service entrance (Figure 3.1), (NEC 230.54[F]). Also, to minimize the wire needed, the meter base should be as close to the drip loop or underground feed as possible.

In most cases, the electrician is responsible for connecting the meter base, the top of which must be 5 feet above ground level, with the roofline where the power company's overhead line connects (NEC 230.50). Overhead lines must maintain a height of 12 feet at the lowest point above residential property and 18 feet above roadways and other right-of-ways (NEC 230.24).

Electricians have a number of options when linking meter bases and overhead lines, including through a gooseneck, through conduit, or through a specialized fitting called a **service head**. **Service entrance (SE) cable** leaves meter-base terminations through a sealed, watertight screw connector and travels vertically to the roofline, where it attaches to the utility line or drip loop. SE cable must be sized according to the service and to NEC 310.15(B)(6).

The 200-amp service requirement of our example must be met with 2/0 Ga copper or 4/0 Ga aluminum (NEC 310.15[B][6]). The layouts of gooseneck and service-head installations are identical except for the 90-plus-degree turn the cable makes at the top from vertical (from the meter base) to slightly below the horizontal entrance to the drip loop (Figure 3.2) (NEC 230.27). The vertical cable run must be secured to solid wood or other material under the siding, if present, using cable clamps no more than 30 inches apart. (NEC 250.5) Double-clamping near the top provides added strain relief. It is good practice to facilitate attachment to the power company by extending the cable 3 feet to 4 feet from the gooseneck or service head.

When using conduit for this type of vertical run, as is typically done to extend service height to the aforementioned 12-foot requirement on one-story houses, run 2-inch conduit or **mast** through a hole cut in the eave of the roof. Extend the cable out the top through a service head as described earlier, and splice 1 or 2 feet

Figure 3.2 Open Cable from Meter Base to Drip Loop—Overhead Service

out of the head. The height above the roof hole should be between 2 feet and 3 feet, and the horizontal distance to the roof edge should not exceed 4 feet (NEC 230.9[A][B]). Additionally, the cable must travel over no part of the roof except the eave, or overhang.

Hints and Tips

The cables used for electrical work are thick, heavy, and hard to bend—harder, even, than many types of conduit. As a result, few electricians can complete cable installations alone. Imagine trying to unroll inch-thick copper cable weighing several pounds per foot while on a ladder. The wire needed for even one story of a home is too heavy to carry along the ground, much less up a ladder. To stay productive and on schedule when completing cable-related tasks, enlist help.

Run galvanized pipe through the hole in the roof from above, down to the meter base. There, thread it into the base and seal it. Then, after pulling the cable through the pipe using pipe or wiring lubricant, cut it, allowing 3 feet of cable for power company connection.

With the cable run complete, attach the vertical SE cable to the appropriate lugs in the meter base. Meter-base lugs are usually giant versions of the lugs in a typical service panel. The holes and securing screws are larger, but the basic termination process is the same.

▶ Connecting the Meter Base

In a properly laid out installation, the connection through the outside wall between the meter base and the service panel is short. When a service panel is more than 5 feet from the service entrance, a master shutoff panel is needed before the breaker panel to disrupt power before it enters the house. When the service panel is close to the service entrance, the panel's master disconnect can serve to cut the residence's power.

Hints and Tips

As in most other areas, the requirements for meter-base connection are driven by local codes. To ensure full compliance, check with local authorities before commencing work.

Hints and Tips

To ease panel-to-meter-base connections (Figure 3.3), leave at least an extra foot of cable on each end as it enters the panel so you have plenty of wire for hookup.

Figure 3.3 Panel-to-Meter-Base Connection

Connect the panel and meter base with SE flat cable rated for water and sun exposure or with rigid/flexible conduit containing thermoplastic high heat resistant nylon-coated (THHN) or thermoplastic high water resistant nylon-coated (THWN) conductors of the appropriate size.

Basement installations can be tricky, expensive, and risky, so they should avoided. When basement panels are unavoidable, an associated **ground fault circuit interrupter (GFCI)** receptacle is required at or near the panel, and the area should be free of obstructions as well as well lighted (NEC 210.52).

Other unique factors apply when a panel is mounted in a basement, something that is often done when a home's first floor starts aboveground. Most locations require protective sleeves, like those used in chimneys, for conductors. When a basement is above ground, a conductor can pass through a sleeve at the basement/first-floor boundary (NEC 230.5). Alternatively, conductors may run through Sched 80 PVC conduit, out the meter base's bottom, and then parallel to the ground, through the basement wall, and into the panel. The humidity of basements, especially unfinished ones, can shorten a panel's lifetime. Basement-mounted panels tend to rust, as do their circuit-breaker **contacts.** In flood-prone areas, float switches are sometimes used to cut power to the panel.

Hints and Tips

When installing a panel in a basement, place incoming connections at the bottom of the panel so that, should water leak into the meter base or conduit, water will not fall into the top of the panel, destroying the main breaker and possibly starting a fire. This is a type of drip loop. Once water fills the conduit, it spills out along the bottom of the panel and falls to the floor before causing issues.

▶ Harnessing the Power of the Panel

Once cable is installed, it is time to attach the **conductors** (black, red, white) to the meter base and panel busses. A service panel fans out one big connection to a finite number of smaller ones. Each circuit is designed to power various household loads. Power is distributed through a panel's four following hardwired busses:

1. *Hot line 1*—Red wire arising from the service entrance, at 120VAC. It connects to a copper-tabbed strip running vertically just to the side of center.
2. *Hot line 2*—Black wire arising through the service entrance, also at 120V. It also connects to a copper-tabbed strip, just to the opposite side of center.
3. *Neutral*—White wire entering through the service entrance. The **neutral** bus is generally just on the inside of the panel.
4. *Ground*—Bus connecting to the grounding system, or ground rod(s), as outlined in Chapter 1.

Note that the ground bus lacks a component exiting to the meter base. Rather, it has a separate conductor that connects the ground bus (also attached to the panel's frame) to Earth ground. At the main service panel, the electrician connects the neutral bus and ground bus so that the residence "agrees" with the power company's definition of "0 volts." This neutral-and-ground connection is optional in the panel itself,

Did You Know?

NEC requires separation of ground and neutral in subpanels per 250.8 and 250.28.

Hints and Tips

Consider split or dual breakers future expansion tools, not primary wiring. That way, customers are equipped to add appliances as they wish.

because a panel often serves as subpanel. In a subpanel, ground and neutral must remain separate. Ground and neutral then combine as one reference only at the main panel.

In addition to busses, a panel usually has a main breaker that is rated for the service. The main breaker, which is up line from the circuit breaker, trips when overall current demand exceeds the panel's rating. The main breaker can serve as a manual disconnect, but this is not its primary purpose.

Select panels based on the number of circuits required for the installation. In general, panels are rated based on two factors:

1. *Maximum panel amperage*—Determined by the service budget and the meter base and service entrance wiring.
2. *Number of supported circuits*—Represented on the panel by two numbers separated by a slash. The first number is the number of full-size breaker slots; the second is the number of circuits available when dual breakers are used. For example, a 40/50 panel could be set up with 40 full-size breakers in 40 slots, or 10 could be configured as dual breakers for 50 total circuits. A rule of thumb for determining the number of supported circuits, which varies by application and local regulation, is one circuit for every five rated amps. Given this, a 200-amp panel would have 40 circuits (200 ÷ 5 = 40).

When mounting a panel, be sure to meet clearance requirements (NEC 110.26[A][2–3][B] through [E]). There must be an access clearance area of 30 inches wide, 36 inches deep, and $6\frac{1}{2}$ feet tall as measured from the floor (Figure 3.4). Additionally, there must be no plumbing above the panel unless a drip tray is installed. Local requirements may vary in this area. As stated before, when a panel is installed in a basement,

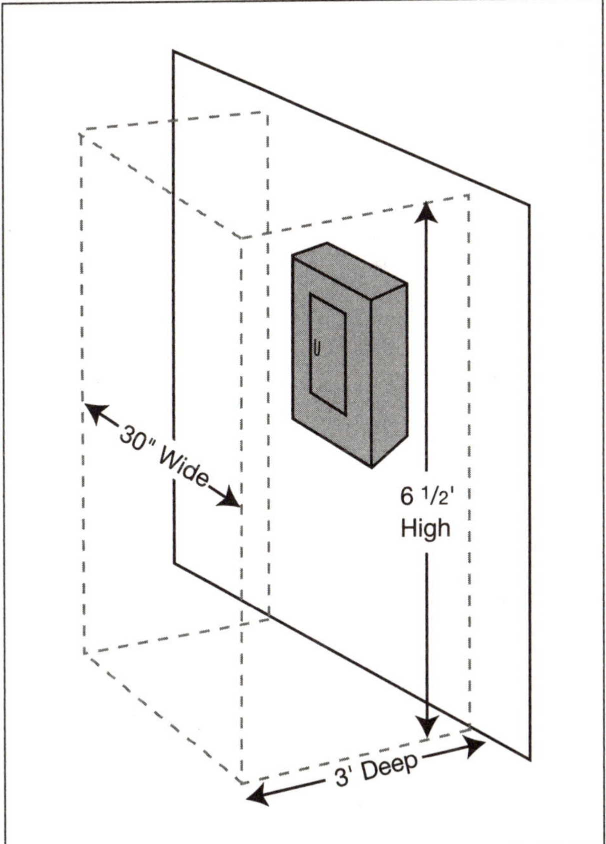

Figure 3.4 Panel Clearance Requirements

a GFCI outlet should reside near the panel to meet NEC requirements.

A panel, like a receptacle, can be **flush** or **recessed mounted.** Flush-mounted panels must adhere to solid surfaces. Drywall is unacceptable. In basements, many electricians attach scrap pieces of wood to walls for panel mounting. The most secure method is recessed between two studs (Figure 3.5), but on 16 inch centers the panel's side entrances are unavailable. As a result, all wiring must enter the panel through top and bottom access holes. Mounting between 24-inches on center studs requires bracing and cross-members, but doing so allows access to at least one side of the panel enclosure. Whatever the configuration, the panel must be

Figure 3.5 Panel Mounting between Two Studs

mounted to the frame on at least two sides. The wall on which the panel is mounted determines which NEC section applies.

In addition to installing wiring from the meter into the panel, the electrician must tap into a ground connection. Recall that a grounding system requires two ground rods buried six feet apart (NEC 250.52[5–8]). Alternatively, two grounding rods that are at least 8 feet long may be encased in concrete under the earth. One end of the grounding conductor, which must be of an approved gauge (NEC 250), is fused or clamped to the rod set using mechanical or heat bonding; the other

Hints and Tips

When clamping to the rod set, use "direct burial," certified clamps. Also, when burying a rod, choose a difficult-to-dig area. The grounding system relies on tight contact between the rod and the earth. Dig the grounding conductor ditch before burying any rods.

Job Connection

After moving into their new home, Chuck and Janet noticed that they tingled when touching their bed's metal foot rail. What started as a curiosity—the pair had assumed they were experiencing static electricity—escalated to a nuisance and then a safety risk. One summer afternoon during a heavy rain, Janet rushed to the bedroom to close a window that had been left open. As she grasped the foot rail of the bed to steady herself, she received a frightening shock. At that point, the couple immediately called an electrician to determine the cause of the problem.

Nelson arrived to find the couple in distress. Janet's right arm was still in pain from the shock, and Chuck insisted on diagnosing the problem before reentering the house. After a short investigation, Nelson invited the couple to the residence's basement. There, he had removed a section of ductwork where it entered the chase to the second floor. Shining his flashlight into the chase, the electrician showed Chuck and Janet a free wire that had stretched across the ductwork and frayed. It was dangling there now, Nelson explained, likely due to furnace vibration.

The bed's foot posts had been resting on the bedroom's register, the coating of which provided resistance and initially gave the couple just a tingle. When water became a factor, however, that resistance declined, leading to shock.

With the couple's authorization, Nelson replaced the faulty wire from the panel to the box, secured it and other chase wires, and reinstalled the ductwork. To address the unbalanced motor and worn mounts of the vibrating furnace, Nelson returned the next day with a new motor and new bushings. To render the system NEC compliant, he added a GFCI receptacle near the breaker box.

ALERT!

To ensure safety, other items in the home, including junction boxes, conduit, water pipes, gas pipes, and any other metal that could conduct electricity, require bonding. Bond any item that poses an electrical risk.

end connects to the panel. For our 200-amp example, NEC calls out 6 AWG. Many electricians replace that option with 4 AWG because of its increased ampacity, which leads to a stiffer ground, and because of local power company requirements.

Grounding electrode conductors are buried between rods and a home's entrance. Because exposure to ground electrode conductors is safe, they must be buried only deep enough so as to avoid inadvertent surfacing or striking by a shovel blade. When there is likelihood of damage, when a conductor exits the ground it should be fastened immediately to an exterior wall and stapled every few feet. Once through the wall, the grounding electrode conductor is bonded to the panel box, usually with a green bonding screw that connects the conductor, ground bus, and panel frame.

▶ Creating Wiring Infrastructure

With the panel installed and connected to the meter base and a grounding system in place, the next step is to create the wiring infrastructure that will deliver energy to all loads and switches.

When new homes are built, electricians usually gain access once outer framing is complete but interior walls are still uncovered. That way, they can easily access walls from both sides, route conductors through floors, and communicate with their teams. Because the structures of existing homes are complete, their wiring jobs tend to be tougher. Electricians must negotiate wiring through covered walls and finished floors and around plumbing and HVAC systems. They must install boxes and fixtures in and on completed surfaces with no chance for error. It is a lot different working on a carefully restored, 200-year-old hardwood floor than on subflooring that will never be seen once the home is complete.

Whatever the structure, wiring jobs involve the five following basic items:

1. *Conductors*—Move electrical charges. A common flat, sheathed, electrical cable, Romex, has two or three conductors and a bare-wire ground.
2. ***Junction boxes***—Used where conductors are spliced (see following) or attached to switches or loads (see following).
3. *Switches*—Control energy flow to a load. Some are on/off, while others are linear, as in a dimmer switch. Timers are examples.
4. *Loads*—Draw on wiring systems. May be things like receptacles, lighting fixtures, and hardwired appliances.
5. *Splices*—Bond conductors.

Following is the procedure for using the preceding elements to install wiring infrastructure:

1. *Install all junction boxes according to plan.* Allot one box for each switch, receptacle, splice, and load. Size boxes correctly to meet fill requirements (NEC 314.16[A]).
2. *Install all required conductors.* As needed, drill holes and notch wood. Note that three- and

four-way switching require extra conductors between switching stations (NEC 300.4).

3. *Terminate all switches, receptacles, and loads, and mount devices* (NEC 314).
4. *Complete all splices* (NEC 314).
5. *Connect conductors to the breaker panel* (NEC 215.3).
6. *Verify and test all circuits* (NEC 220).
7. *Install trim plates and a panel cover* (NEC 110.1[E]).

▶ Understanding Junction Boxes

Junction boxes come in many varieties. Metal boxes, for example, have removable sides so they can be grouped two or three at a time for added volume. In fact, metal boxes are required when flush-mounting to a wall or in commercial installations. Metal boxes used to be the standard, but their plastic counterparts are now gaining wider acceptance because they are cheaper. Some plastic boxes nail in place, while others screw in and are therefore removable.

For the electrician, the biggest challenge with junction boxes is choosing one in an appropriate size, a size that accommodates all needed wires and **wire nuts.** More than simply answering a question like, "Two switches or three?" this "conductor fill" exercise involves relating a box's cubic inch volume, which is stamped in the box's interior, with both the number of conductors entering the box and the loads the box sources, if any.

Consider the following conductor fill requirements:

Conductor	Cubic inches of box space
#14	2
#12	$2\frac{1}{4}$
#8	3
#6	5

In addition, consider that all grounds count as one conductor of the common size and that a receptacle or switch counts as two conductors of the common size. Given that, assume that speed control for a whole-house attic fan is powered by 8 Ga wire mounted in a junction box between two studs. There would be seven total 8 Ga conductors (four [2 sheaths, 2 conductors each, or 2 + 2] plus one ground, plus two for the switch [4 + 1 + 2 = 7]).

From the preceding table, 3 cubic inches per conductor times seven conductors is 21 cubic inches. At just over 22 cubic inches, one gang box of the maximum available depth ($3\frac{1}{2}$ inches) would be an appropriate choice (NEC 314.6).

Now assume you have a three-gang box (Figure 3.6) with two switches. The RH switch (center position in the box) controls the bottom half of a split receptacle **strap** in the third space, the top half of which is always on. The LH switch is part of a three-way switch setup.

To determine the number of conductors for each switch, do the following:

Switch 1 (three-way switch)

1 sheath of 14-2 (2 conductors),
1 sheath of 14-3 (3 conductors),
and switch (2)

2 + 3 + 2 = 7 total conductors

Switch 2 (split receptacle control)

2 sheaths of 14-2 (4) and switch (2)

4 + 2 = 6 total conductors

To calculate the number of conductors for each outlet, do the following:

Outlet 1 (switched; must be tabulated separately because the receptacles are split, meaning they are sourced

Figure 3.6 Three Gang Junction Box

by two separate circuits, and the connecting tab(s) on the receptacle strap are broken to isolate the two)

1 sheath 14-2 (2) and receptacle (2)

2 + 2 = 4 total conductors

Outlet 2 (unswitched)

2 sheaths 14-2 (4) and receptacle (2)

4 + 2 = 6 total conductors

plus, 1 conductor for ground.

Therefore, the grand total is 24 conductors (7 + 6 + 4 + 6 + 1). Using the rule that requires 2 cubic inches per conductor, this installation requires 48 cubic inches.

With the preferred junction box sized correctly, the electrician must either mount that box in an existing wall or affix it to studs in an unfinished one. "**Old work**" boxes, boxes intended for existing structures, are designed for installation in drywall or on brick, block, or stone surfaces. A metal box must be flush-mounted on the surface with conduit sourcing through

Hints and Tips

Especially with plastic junction boxes, remove punchouts before box installation (NEC 314.19–20).

Hints and Tips
To avoid having to measure every box intended for installation at the same height, devise a jig out of scrap lumber for reference.

Did You Know?

When installed around a wooden framework, Romex is stapled (NEC 334.30). In commercial work, conduit attaches to metal studs (NEC 342.30).

the wall or along the wall surface (NEC 314.21,22 and 23[1] and [2]). For new construction there are labor-saving box styles, or "**new work**" boxes, that can be nailed to studs quickly and firmly, with the right overhang for different thicknesses of drywall.

Before mounting any junction boxes, electricians should consult their plans to determine how conductors will enter those boxes. Each box has cutouts, or **punchouts**, that must be removed to accommodate wire sheaths in and out. Once each box is installed, the electrician examines it for cracks or splits and replaces it when it has lost its integrity.

▶ Running Interconnecting Wires

Once all junction boxes of the correct capacity are installed, it is time for the electrician to run wire between those boxes and the previously installed panel. From the panel, wire runs vertically and horizontally until reaching the termination point. The wiring jobs of the most successful, experienced electricians are neat. Novices, in contrast, may be inclined to bundle all wires from the panel into one tight channel and to split off what is needed as the wires route through the residence. The bundling approach often works with low-voltage wiring, like that for computers, but it fails to work with high-voltage wiring because high-voltage wiring generates heat.

It is necessary to derate wiring's ampacity based on the wiring's temperature (NEC 310.15[B][2]). Hot areas require thicker wire for the same amount of current run through cooler ones. Therefore, it is best to

Hints and Tips
To save time and labor, use a wiring spool. Use Figure 3.7 and scrap lumber to create one. Once the spool is cut, slip the wire over the shaft.

Figure 3.7 Spool Template

Did You Know?

When installed properly, wire always runs straight up or straight across, never diagonally. All wire angles are 90 degrees. This gives electrical work the appearance of neatness and planning.

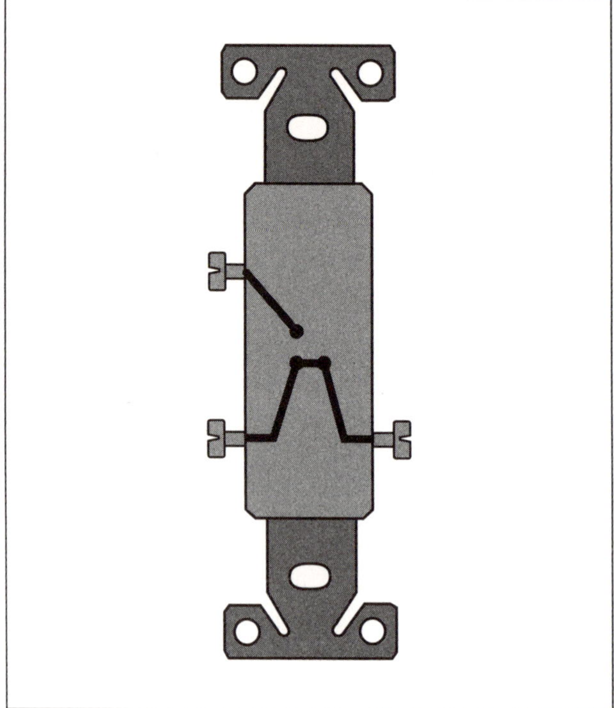

Figure 3.8 Three-Way Switch Schematic

limit Romex connections to groups of three or fewer (NEC Table 310.15B2a).

Wiring jobs in typical frame construction require notches, holes, or both to accommodate three or fewer wires. When cut improperly, those notches and holes can weaken a home's frame, compromising the structure. In addition, nails or screws can puncture wires run through those openings, especially after walls are finished. To address these concerns, NEC requires removal of no more than 40% of a frame piece, measured horizontally (NEC 422.17). That limits a stud's wiring hole to about $2\frac{1}{4}$ inches. It also limits the depth of any notch on a stud face (NEC 300.4[A][1] and [2]). NEC 422.17 requires nail guards for notching and wall-face-to-hole-edge distances under $1\frac{1}{4}$ inches.

To safeguard wiring in attics, electricians cannot bridge joists with sheaths alone (NEC 320.15). Trays, wire tracks, or bridges are required when running Romex across joists.

▶ Installing Three- and Four-Way Switches

By design, three- and four-way switches running Romex require extra conductors, called **travelers**, between them and the load. The basic three-way switch (Figure 3.8) provides switching for one load at two locations. Those locations for a stairway landing light, for example, are upstairs and downstairs.

Toggling the switch at either location turns its load on or off, independent of the other switch. Such a switch, therefore, has no "on" or "off" indicator. In Figure 3.8, 12-4 or 14-2 wire runs to the first switch box, 12-3 or 14-3 runs between the first box and the load, and 12-3 or 14-3 runs between the load and the second box. As good practice, and as required by NEC 200.7, travelers should be taped to indicate their function. For example, because the white wire from the load to the second switch is switched hot, it should be wrapped with black electrical tape to indicate that it is not a neutral line. Figures 3.9 and 3.10 are sample three-way-switch wiring diagrams.

Four-way switching, which requires two three-way switches as well as one four-way one, is a more com-

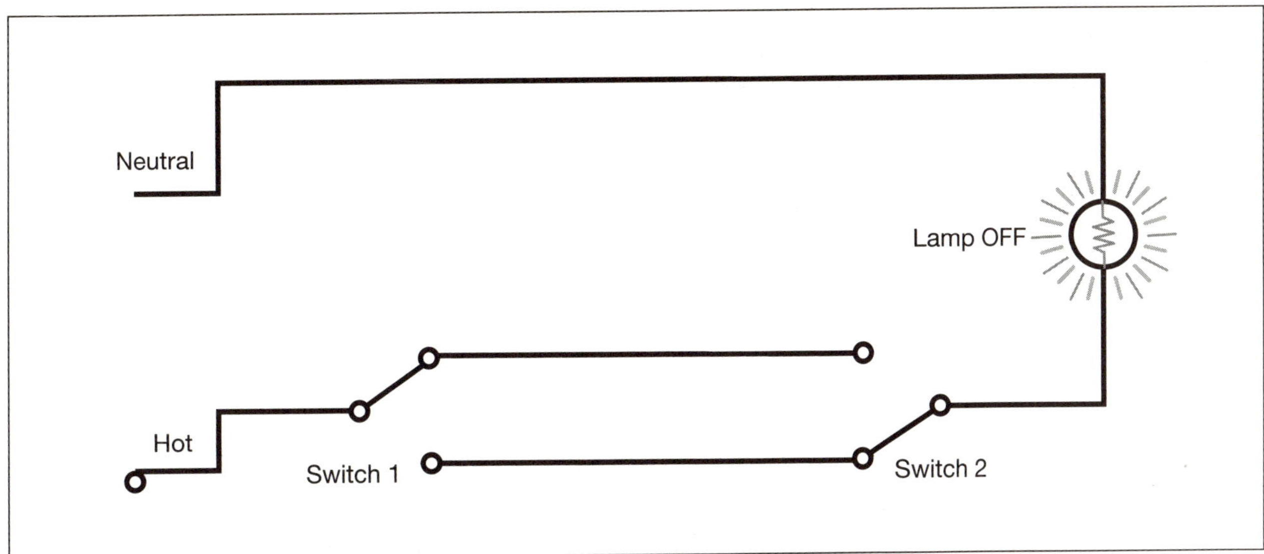

Figure 3.9 Three-Way Switch Wiring Diagram

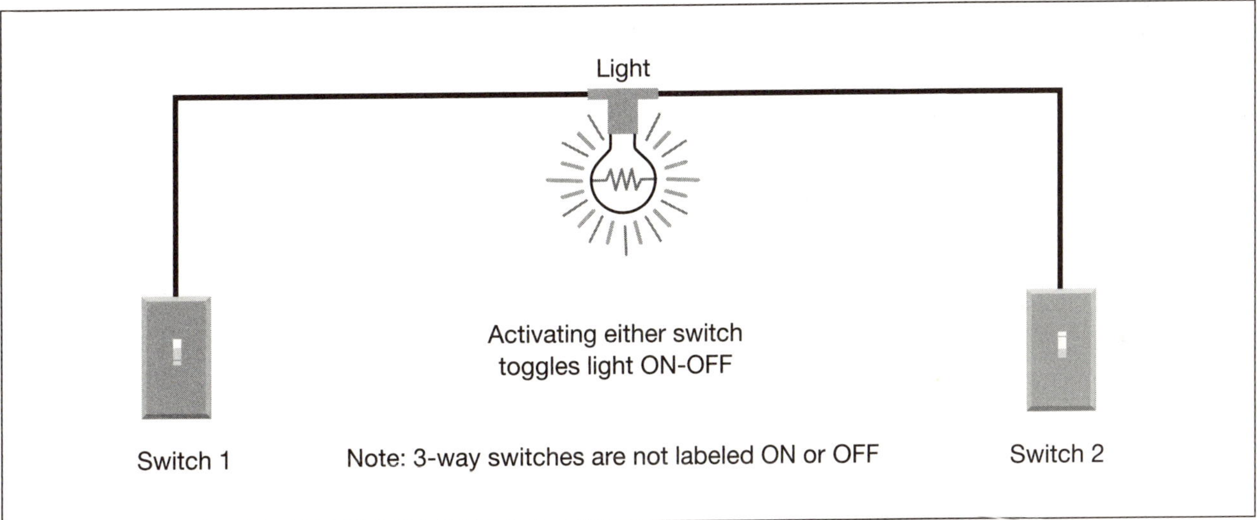

Figure 3.10 Three-Way Switch Circuit

plicated approach (Figure 3.11). In four-way switching, a load switches on or off from three locations. Like three-way switches, four-way switches lack on/off indicators. When a load is off, moving any of the three switches to the opposite position energizes the load.

As with three-way switching, travelers that do not follow color code require black tape (NEC 200). For example, some white wires, usually at 0 volts, that function as "switched hot," require tape indicating that they are not ground.

When electricians run conductors to and between the junction boxes they have installed, they typically leave 3 feet of extra wire from the point where the wire enters the panel. At junction boxes, they typically leave

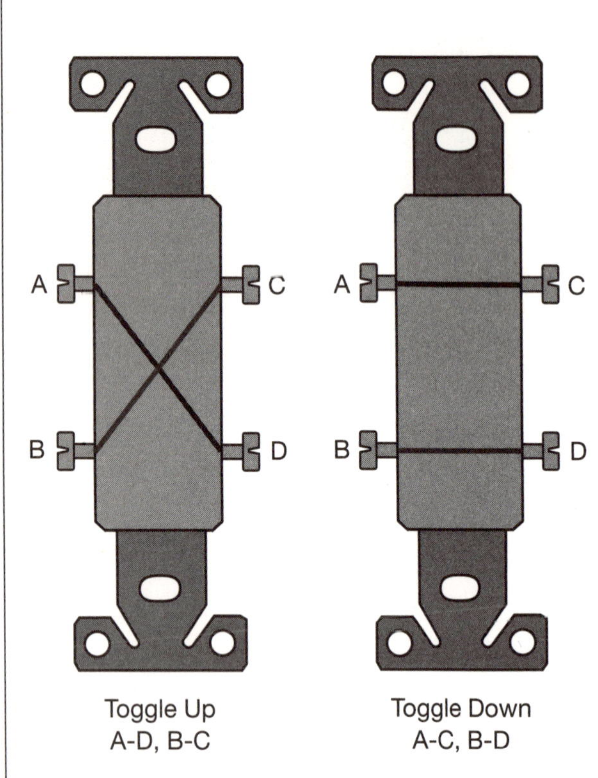

Figure 3.11 Four-Way Switch Schematic

Figure 3.12 Dryer Wiring on 240V Circuit

an extra foot to dress cables, splices, and terminations properly in breaker boxes and junction boxes alike.

For 240V circuits in a home, NEC now requires four-conductor cable (one per phase, one neutral, one equipment ground for use with circuit boards). Local requirements dictate the use of flex conduit, or Greenfield, for encasing wiring from boxes to appliances in 240V installations. Figure 3.12 shows wiring for a typical 240V clothes dryer. Figure 3.13 reveals wiring for stove and oven combo units requiring 120V and 240V.

When it comes time to install receptacles and switches, there are several quality levels, or grades, from which to choose. Following are the basic options:

- *Residential*—Cheapest option. Residential components are lightweight, brittle devices with thin plastic and metal conductors that can crack or break easily and that have a very low tolerance for current overload or even supplying rated current for an extended period.
- *Commercial, heavy-duty, industrial*—Mid-grade offering. These options offer more robust construction than residential choices and are more tolerant of extended use and vibration. They also better address environmental concerns.
- *Hospital-grade*—Highest cost and quality option. Hospital-grade elements provide the reliability and quality needed to sustain life. Receptacles of this level are required in patient areas where occupants are left unattended to sleep or recuperate.

Given cost considerations in new construction, the tendency is to meet, but not exceed, code requirements. Hospital-grade receptacles may be installed in a residence, but they are not required. As with most things,

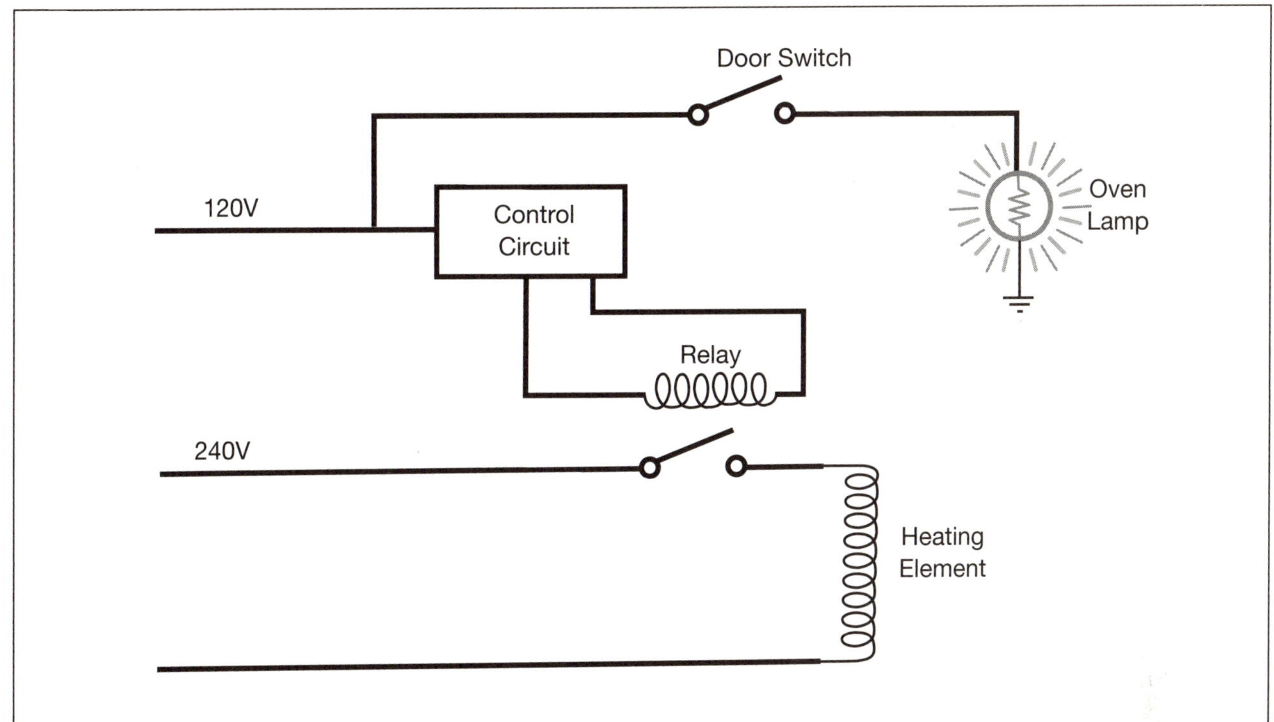

Figure 3.13 Hybrid 240/120V Wiring for Oven–Range Combo Unit

> **Hints and Tips**
> Cost is a major factor in electrical jobs. In general, successful bidders submit the lowest prices for jobs, which often means they must cut corners wherever possible. The best electricians, however, find ways to save money that do not erode their quality of service. One of the easiest cost-cutting measures is to purchase supplies in large quantities. That way, the electrician can install higher-quality materials while balancing the need for budget vigilance. Particularly for electricians who freelance, success stems from the repeat business of satisfied customers. Install the highest-quality components that you can afford, and always deliver top-of-the-line service.

there is a middle ground. The most inexpensive receptacles are less safe and effective than their hospital-grade counterparts.

Receptacles are available in 15-amp and 20-amp current ratings (Figure 3.14). Plugs that carry 15-amp loads, standard versions that have two parallel prongs,

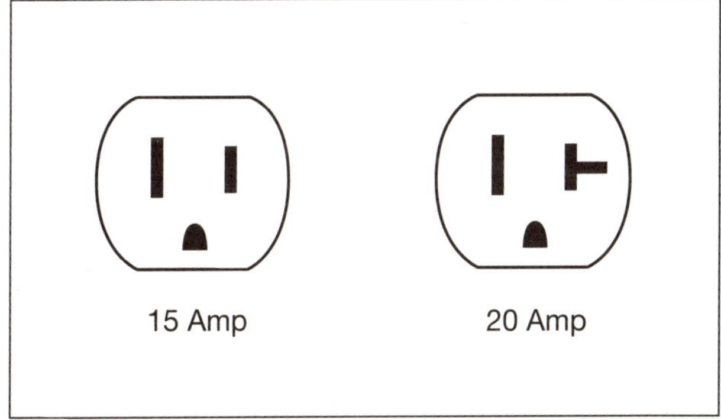

Figure 3.14 Receptacles: 15-Amp and 20-Amp

ALERT!

When working on existing, occupied residences, astute electricians always look for discrepancies in circuit capacity and outlet configuration because such discrepancies represent safety hazards as well as code violations. When a 20-amp outlet draws on a 15-amp circuit, for example, offer to replace the 20-amp outlet with a 15-amp one to align circuit ampacity or, if it is within service limits, replace the wiring and breaker to bring the circuit in line with the outlet. Should the customer choose to exercise neither option, at least red-tag the installation to highlight the risk.

plug safely into 20-amp receptacles (NEC 210.21–53). Per National Electrical Manufacturers Association (NEMA) standards, the 20-amp receptacle has an extra horizontal slot that connects a 20-amp rated load, which has one vertical prong and one horizontal one. These types of plugs cannot safely plug into 15-amp receptacles.

Hints and Tips
Before installing a receptacle into a junction box, wrap electrical tape around the receptacle to protect the terminations from inadvertently contacting the box or other objects.

Hints and Tips
In areas requiring it, it is often more convenient to provide GFCI protection at receptacles than at panels. Unlike in standard receptacles, the line and load sides of a GFCI receptacle must be oriented correctly for the receptacle to work properly. Therefore, when wiring a junction box to accept a GFCI receptacle, know which side is which and route wires through the correct knockouts to minimize wire crossing in the box.

In addition to varying in current rating, receptacles are available in single (simplex) or double (duplex) configurations to energize one or two devices, respectively. Usually, duplex receptacles are installed in homes, but simplex receptacles have some specific uses. In basements, for example, a freezer requires either a GFCI duplex receptacle or a non-GFCI simplex receptacle. While GFCIs may be safer than other alternatives, they are prone to false tripping, which can be problematic when a freezer is full of food. As a result, NEC mandates GFCI receptacles in unfinished basements (NEC 210.52).

▶ Exploring Switch Basics

Like receptacles, **switches** are available in grades, but 15 is generally the maximum ampacity for residential switches. Residential-grade switches are thin, shallow, and easy to mount in junction boxes, but they are brittle, unreliable, and unable to sustain rated current. For their part, heavy-duty switches take more box space and are more difficult to work with.

Hints and Tips
Do not use the push-in terminals of the less expensive residential-grade devices, because they wear down, become intermittent, and fail. As a result, they compromise safety.

Did You Know?

Dimmers and other rheostatic devices are wired like regular switches. Their only difference is in the switching actuator, which is independent of wiring approach.

In addition to grade and current ratings, switches are defined by the number of **poles**, or conductors, that can be switched. A one-pole switch can control one load and breaks/makes one connection (the "hot") to that load. However, power should pass through the switch before it reaches the load, not after, because if it is switched after, voltage is still at the device and vulnerable to trying to find a way to ground. Double-pole, single-throw switches control two independent circuits. They switch 240V loads on and off. A 240V system has two independent hot lines, and both must be switched using the same actuator (NEC 404).

As discussed previously in the wiring section, three- and four-way switches are specialized units with extra terminals requiring extra conductors. Correct termination of these switches requires an understanding of their unique layouts. A three-way switch has three terminals: one marked COM, for "common," and two others that are unmarked. Unmarked terminals are for traveler wires. A four-way switch, in contrast to a three-way one, has two sets of terminals for two sets of travelers.

▶ Identifying Proper Fixture Uses

Lighting fixtures and appliances come in various stages of readiness for installation. Some arrive with **pigtails**, or fixture whips, which are preattached wires ready to be spliced into a circuit, while others lack all kinds of wiring. The extra wire left at junction box installation helps accommodate fixtures like the latter. Ceiling-mounted fixtures may require reinforced mounting through mechanical means. Sometimes, the hardware these fixtures come with affords too little support. Fixtures weighing more than 50 pounds require supplemental mounts (NEC 701[A], 314.27[A], [B], and [D]). Cross-members between joists are one example.

Hints and Tips
Remember that three-conductor cable is required between switches and loads. Use black tape on white traveler wire so the wire is not assumed to be grounded.

Hints and Tips
When buying or specifying ceiling fixtures exceeding the weight limit, look for optional mounting kits with all the parts needed for safe and secure installation.

▶ Applying Outside Fixtures, Boxes, and Switches

Inside and outside wiring procedures are identical except in two cases. First, outside receptacle wiring must be GFCI protected, preferably at the breaker (NEC 210.8[B][3]). Second, outside wiring must run behind a water barrier, usually a gasket or another rubber barrier at the service opening to the fixture (NEC Article 100, definition of *waterproof*).

Did You Know?

Wire nut packages indicate how many conductors can be spliced by the wire nuts inside. For example, a given wire nut may be able to splice three 14 Ga conductors or two 12 Ga ones.

▶ Terminating Loads, Switches, and Receptacles

Once all wiring is roughed in (i.e., wires are run but not terminated) and devices are identified and purchased, the electrician attaches the wires to the loads and switches in their respective junction boxes and completes any required wire splices. **Splices** (wire nuts) and terminals are two options for these connections.

When the splice or wire nut option is exercised (Figure 3.15), the wires are first stripped, twisted together tightly, and cut. Then, a wire nut is screwed onto bare conductors to secure connections and safeguard

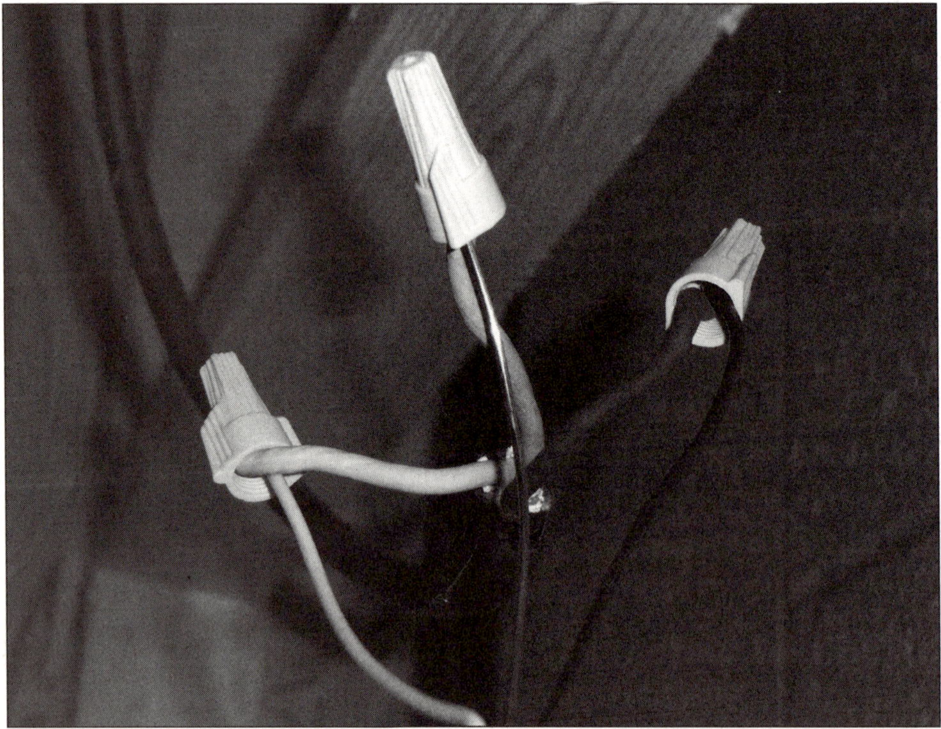

Figure 3.15 Wire Splice or Wire Nut

against shorts. When terminals are used (Figure 3.16), wires are stripped to length, the wire is looped, and the loop is placed over the shaft of the partially removed screw terminal where the screw is tightened to secure the conductor to the switch or receptacle.

When connecting components through splicing and wire bonding to screws, tuck the wires back into the boxes securely and mount the devices to their respective boxes. Leave trim plates off initially to facilitate readings or troubleshooting if needed.

Hints and Tips

Have various types of wire nut types on hand during installations, and be sure to use correctly sized wire nuts for applications. Check a splice's integrity by attempting to pull the wires out of the wire nut. Inspect the splice to ensure no bare wire is exposed at the bottom of the wire nut, which could short to the case or to another conductor in the box.

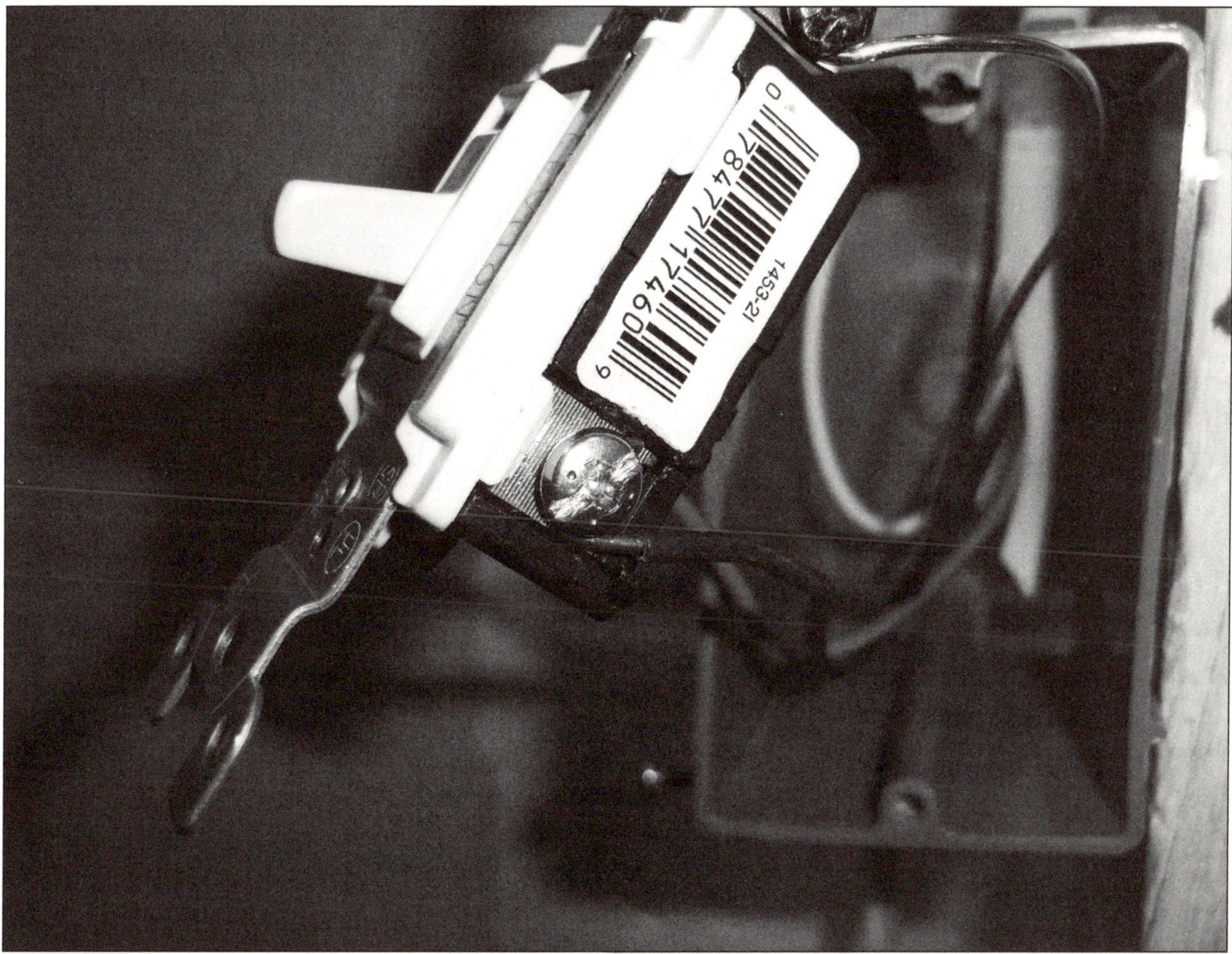

Figure 3.16 Wire-to-Screw Terminal on a Switch or Receptacle

ALERT!

Many inexpensive switches and receptacles, designed for do-it-yourselfers, feature small holes into which wires can be pushed to secure terminations (Figure 3.17). Do *not* use this type of termination. It is less secure than a screw-down terminal and can be dangerous when the retaining spring inside the switch, which cannot be inspected, proves defective.

▶ Routing and Dressing Cable

With all loads, outlets, and switches wired, the electrician returns to the panel. Effective electricians plan their installations in writing, documenting before work starts how circuits are arranged in a panel. They also number circuits using the panel's circuit breaker slot numbers. By the time panels are in place and all circuits are wired, these electricians know exactly how many circuits a residence has, as well as where those circuits go and where they connect to the panel. Recall that wiring involves two hot circuits sharing a common neutral. The goal is to balance those two hot sources.

Figure 3.17 Friction-Type Terminals on a Switch

Consistency is the key to quality electrical work when finishing wiring to a panel. First, cross over no wires in a panel. All wiring terminating on the right side of the panel should enter the panel from the right side; all wiring to be attached left should enter left (Figure 3.18). Keep the wiring toward the peripheries of the

Did You Know?

Most panels feature odd-numbered circuits on one side, even on the other.

Figure 3.18 Proper Dressing of Panel Wiring

panel box to avoid breaker operation interference and pinching during panel cover reinstallation.

As a second step toward quality electrical work, cut conductors to the proper length by first keeping them vertical along the edge of the panel interior, then into a 90-degree bend to the appropriate breaker. Conductors should terminate into the bus with no bare wire visible as the conductor enters the bus, and bus terminations should be made close to the breaker carrying the protected conductor (NEC Article 408) (Figure 3.19). In correctly wired panels, it is easy to trace a sheath from its source at the breaker and busses to its exit from the panel.

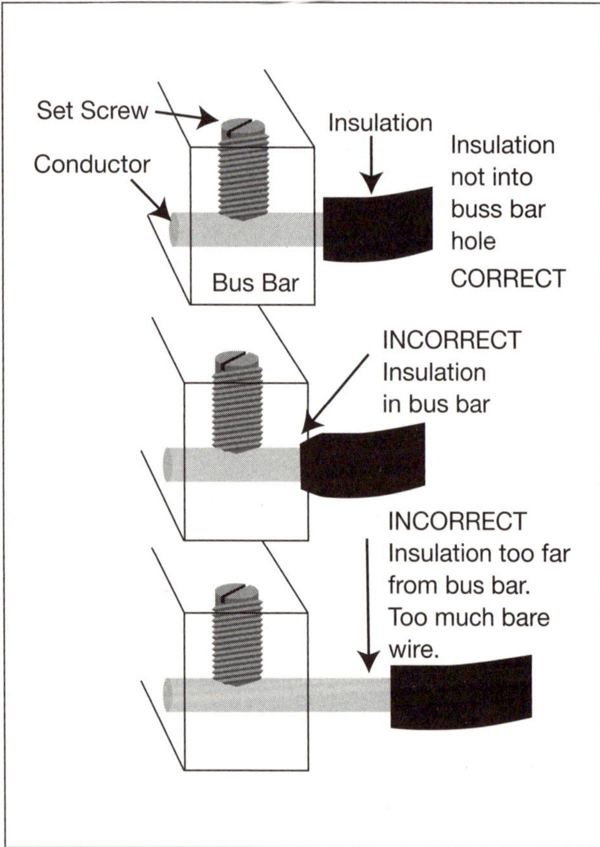

Figure 3.19 Proper Termination of Panel Wiring

▶ Conducting Testing and Verification

With all the preceding tasks concluded, the installation might appear complete. The service entrance is wired, the meter base connects to the panel, the panel connects to the switches and receptacles, and fixture and appliance hookups are in place anticipating the equipment's arrival. Testing and verification are final steps.

Testing can be done under one of two conditions, depending on whether the power company has connected power to the meter base yet. Static testing occurs without power applied to the system. It establishes only resistance, or continuity, between points. With this type of testing, electricians are charged with checking to ensure that the following exist:

- High resistance between each hot and neutral line.
- High resistance between hot and hot. The two separate hots are not to be connected.
- Zero ohms between neutral and ground.
- Zero ohms between panel ground and the grounding rod set.
- Grounding of HVAC ductwork, plumbing (when installed and grounded), and any other grounds made while developing system ground.

Hints and Tips

In anticipation of appliance delivery, make sure that any stubs intended for subsequent termination are secured and that no bare wires are protruding.

Job Connection

Most homeowners have experienced dead circuits. The same is true of most electricians. Troubleshooting a home's electrical system can be very time consuming and laborious, especially, for example, if the problem is on the second floor, and the panel is in the basement. Yelena learned this the hard way when she was hired to solve a recurring, intermittent shutdown of a patio post light.

When Yelena arrived onsite to find the post light working, she decided to ask her journeyman, Chuong, to determine which circuit fed the light while she inspected the fixture directly. Unfortunately, the breaker box held no clue as to which wire fed the outside implements. When Yelena shouted inside, "Why's the light still on?" Chuong began turning off breakers. Fortunately, he hit the correct breaker on only his second try. However, the breaker felt mushy, not crisp as it should be, so he replaced it after consulting with Yelena. Believing they had solved the problem, the pair left a note for the homeowner and departed.

The next day, Chuong received a telephone call from the homeowner, who was enraged. The homeowner's son had been asleep in an upstairs bedroom while the electricians worked. When Yelena and Chuong switched the power back on, the son's alarm clock failed to repower, causing him to miss his flight to an out-of-town job interview.

From this experience, Yelena and Chuong learned some valuable lessons. First, when the inside door with the circuit directory has a label, remove it (NEC 408.4). Such labels are too small, and their entries are either incorrect, illegible, or incomplete after circuits are changed or added. Second, label each incoming wire sheath with cloth tape and indelible ink to indicate its breaker number. This should happen at both the panel end and at the fixture/outlet/switch end.

Of course, Yelena and Chuong learned other things, including that some panels have slots for small slips of paper that can carry circuit descriptions for each breaker. The pair learned to use those slips and to leave space for additional writing. They even decided to leave additional slips and to carry extras with them. For each breaker, the pair learned to apply labels with breaker numbers using cloth tape and a label maker.

The final lesson for the two electricians involved receptacles. They learned that each receptacle and switch in a home should have a label under the trim plate indicating which circuit it is on. Similarly, each hardwired appliance should have a circuit number printed inside the box, where it can be easily seen. While not an NEC requirement, such practices can save time and prevent unnecessary service interruptions.

Hints and Tips

During dry weather, which is when problems are likely to occur, use a **megger**, a piece of test equipment that measures resistance to ground, to measure the effectiveness of a ground plane. If issues arise, begin by disconnecting circuits to locate the problem. Do not rush to energize the system, because these problems will persist when voltage is applied.

Once a system passes all static tests, it is time to start dynamic testing. **Dynamic testing**, which entails making measurements with the system powered up, involves the following steps:

1. Before the power company turns on the power to the demarcation line, shut off all breakers, including the main one.
2. Verify that power is on by measuring the two hot voltages to neutral at the main disconnect.
3. Energize the disconnect, and examine the electric meter, which should be immobile. If it is moving, de-energize the disconnect and inspect the meter again. If the meter has stopped turning, a leakage exists between the main disconnect and the circuit breaker hot side. Look for arcs, and use an infrared remote thermometer to look for panel hot spots. If the meter continues to turn after the mains are disconnected, the problem lies with the utility company or between the meter base and the service entrance.
4. When it is established that there is no current leakage, place the clamp lead of the voltmeter around the Romex leading from Breaker 1. Verify that 0 amps are present, and then energize Breaker 1. Verify again that no current is flowing. If current is flowing or if the breaker trips, investigate. Examine the circuit to determine if a lamp is burning or if something is plugged in. If they are not, find the leak in the wiring and repair it before moving to Breaker 2. Repeat these steps with each breaker, turning each one on and off in succession until all are tested and troubleshot.
5. Turn all circuit breakers on, and measure the total current coming in through the service entrance. If it is not 0 amperes, locate the source of the current draw by repeating the breaker testing. If it is 0 amperes, use a circuit tester to check the polarities of all receptacles and fixture pigtails and verify their operation. Rewire as needed to achieve the correct polarity.
6. At the breaker panel, compare panel voltages to ground and to the ground connections established and static-tested earlier. Ensure they are all at 0 potential before shutting off the main disconnect.
7. Install the panel cover, as well as trim covers for all outlets and switches. Finalize any fixture installations based on construction progress. For example, if drywall is not yet installed, simply lean the switch and outlet trim plates against the boxes they are to cover. If floor molding was removed to complete the installation, reinstall it.
8. Make one final check by reenergizing the system and measuring total current and ground potential.

▶Chapter Review

1. _____________ conduit is rated for a service over a roofline.

2. PVC pipe is appropriate for overhead electrical service.
   - **a.** true
   - **b.** false

3. When installing a subpanel in a 200-amp service, _____________ amperes are allowable.
   - **a.** 10
   - **b.** 50
   - **c.** 100
   - **d.** 200

4. Why does an electric dryer need a neutral?

   ____

   ____

   ____

   ____

5. The smallest service for a residence is _____________ amp.
   - **a.** 100
   - **b.** 120
   - **c.** 240
   - **d.** 400

6. A double-pole breaker is used for which of the following voltages?
   - **a.** 100
   - **b.** 120
   - **c.** 200
   - **d.** 240

7. Which kitchen receptacles require no GFCI protection?

   ____

   ____

   ____

   ____

8. There is a limit on light switches.
   - **a.** true
   - **b.** false

9. What is the benefit of using 240V over 120V?

   ____

   ____

   ____

   ____

10. In a garage, receptacles should be __________ high off the floor.

11. In a bedroom circuit, 12-2 Romex is appropriate.
    - **a.** true
    - **b.** false

12. A detached garage requires electrical service.
  a. true
  b. false

13. Why do all outside receptacles require bubble covers?

____

____

____

____

14. Name three places in a home where split receptacles are properly installed.

____

____

____

____

15. From a doorway, the first receptacle should be no farther than _____________ feet.

16. What size ground is needed for 10-2 wire?

____

____

17. The cubic inches of a device box is available only by calculation.
  a. true
  b. false

18. Romex should be stapled _____________ apart.

19. How far from a box is a first wire stapled?
  a. 4 in.
  b. 6 in.
  c. 1 ft.
  d. 4 ft.

20. NEC allows installation of a breaker in a panel box that differs from the manufacturer's panel.
  a. true
  b. false

21. The bare wire arising from the power company side is considered ______________________.

22. Most local codes allow electricians to use wire ties to strap wire to a three-gang box.
  a. true
  b. false

23. How many wires go on a three-way switch?
  a. 1
  b. 2
  c. 3
  d. 4

24. What is wire mold commonly used for?

____

____

____

____

25. In general, ______________ percent of a breaker's rating should be used.

26. All metal boxes require bonding with a ground screw and clip.

   a. true
   b. false

27. On a dedicated individual branch circuit using a 20-amp breaker and 12 AWG wire, a ______________-amp receptacle appropriate.

   a. 5
   b. 10
   c. 15
   d. 20

28. Identify four dedicated circuits in a residence.

____

____

____

____

29. In a typical installation, ______________ degrees of bend are allowed in conduit before a pull box is needed.

30. What size ground wire is used with the following sizes of nonmetallic sheathed cable?

14 AWG ______________________________

12 AWG ______________________________

10 AWG ______________________________

8 AWG ______________________________

CHAPTER

Protection in the Home

CHAPTER SUMMARY

Lethal even in minute amounts, electricity is an unavoidable potential hazard of residential installations. Electricians safeguard their customers and those customers' residences by remaining vigilant about protective measures throughout the wiring process. When installed and replaced properly, fuses and circuit breakers are devices that help maintain overall safety. GFCI and AFCI provide an added layer of protection. Smoke and other detectors approach protection from a high-level perspective. To deliver the highest level of safety before, during, and after a home installation, the electrician's personal safety must also remain paramount. Personal protective equipment, when paired with a proper respect for electricity, allow installations to proceed without serious incident.

▶ Learning Objectives

- Define the term *protection*
- Compare fuses and circuit breakers
- Identify the GFCI circuit and state its application
- Characterize the AFCI circuit and its application
- Explain how to install a smoke detector
- Outline methods for self-protection in the electrical trade

▶ Key Terms

arc fault
arc fault circuit interrupter (AFCI)
bare-wire ground
electrode
fusible link
smoke detector

▶ Understanding the Scope of Protection

Electricity is a form of energy that, even in small amounts, can be lethal. The fact that seasoned electricians risk shock or electrocution underscores the risk to the relatively uninformed general public. Because electricity poses such weighty hazards, a significant portion of the NEC is devoted to electrical protection. The largest NEC article, 250, is devoted to grounding, for example, while Chapter 2 is devoted to protection overall. Simply put, *protection* is safety, the survival of customers and business. Through training, protective devices, and proper NEC application, electricians safeguard their customers and those customers' residences.

In theory, all electrical conductors offer protection because they are metal. That protection is limited, however, because every metal has a melting point and current is proportional to heat. Given enough current, any conductor will melt, breaking the circuit.

Where in the building that conductor melting occurs is the concern. If it were to occur near a kitchen cabinet full of paper towels, for example, the results could be disastrous. Electricians can provide protection by installing devices in safe places that melt at predictable current levels. Those devices are called fuses.

Did You Know?

The fuse is the original protective device in household circuitry. In most modern homes, however, circuit breakers have replaced fuses. Whatever the protective device, melting point is measured in amperes.

▶ Appreciating the Value of Fuses

A fuse has two **electrodes**, or terminals, joined by a fusible link. A **fusible link** is a thin piece of metal that has a low melting point. A fusible link's current rating is determined by its cross-sectional area. That way, depending on its diameter, one type of metal can accommodate a wide range of currents when in fusible-link form.

Figure 4.1 Barrel and Screw-In Fuses

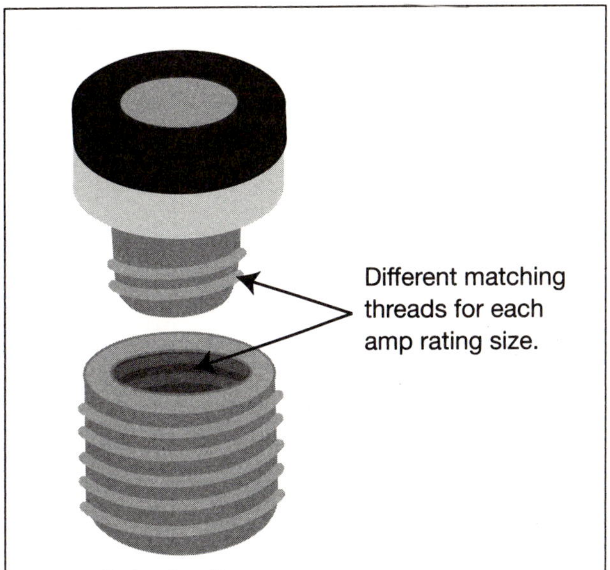

Different matching threads for each amp rating size.

Rejection Base (Tamper Proof)

Figure 4.2 Type S Fuse

The resistance between a fuse's two terminals should be 0 ohms. Once a fuse has blown, it is best discarded because it lacks value. An ohmmeter is the tool used for testing fuse terminals.

Fuses come in two basic styles: barrel and screw-in (Figure 4.1). The windows in screw-in types reveal whether the fuses' fusible links have melted (i.e., the fuses have "blown"). Old-style fuses were interchangeable between ratings, so it was easy to install fuses incorrectly. Newer fuse types, like Type S (Figure 4.2), offer different threads for different fuse ratings so mismatches cannot occur.

Mismatches are just one problem plaguing fuses over time. Fuses have also traditionally worn easily, increasing expenses. For motor-run appliances, time-delay fuses are often used. Motors are unique because, unlike other loads, they require more current at startup than they do at full, running speed. A motor that requires 5 amperes to run, for example, may require 9 or 10 amperes to start. A 5-amp fuse will break the

ALERT!

Always replace fuses with fuses of the same types and ratings. Failing to do so risks overheating, and possible fire.

Hints and Tips

On the door of a fuse panel, note each fuse's type and rating. In addition, advise customers to keep spare fuses on hand. To enhance customer service, consider keeping fuses on hand to bestow on needy customers gratis.

circuit at startup, while a 10-amp fuse will protect neither the motor nor the circuit in the event of an overcurrent following startup. The solution is a time-delay 5-amp fuse, which delays protection until after the motor has started and stabilized.

> **ALERT!**
>
> As a stopgap measure, some homeowners replace blown fuses with pennies. In addition to illustrating a lack of wiring knowledge, this practice represents a serious safety hazard. When confronted with a situation like this, take the opportunity to educate customers. Discuss the dangers of bypassing protection and offer to upgrade the customers' systems to circuit breakers, which are simpler because they are resettable and resemble switches. Oftentimes, a fuse panel is a sign of obsolete service.

▶ Opting to Use Circuit Breakers

Given the challenges inherent in fuses, circuit breakers are easier overall to work with. In addition to being resettable, they enjoy longer service lives. While circuit breakers, like fuses, generate disconnects through excess heat, they look and work more like switches. In essence, a circuit breaker is a switch that is controlled by temperature rather than human intervention.

Like switches, circuit breakers come in single-pole and dual-pole configurations (Figure 4.3). Single-pole switches and circuit breakers are used for 120V systems; the double-pole versions of each apply to 240V. Occasionally, breakers are ganged in one case to create a dual breaker. At other times, they are miniaturized so

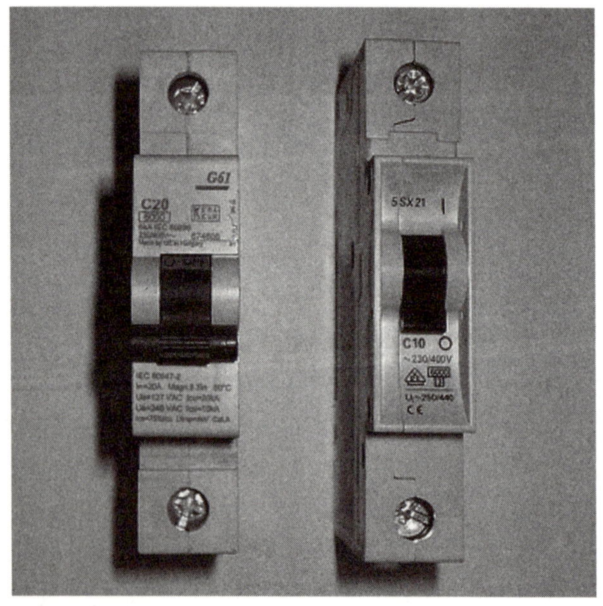

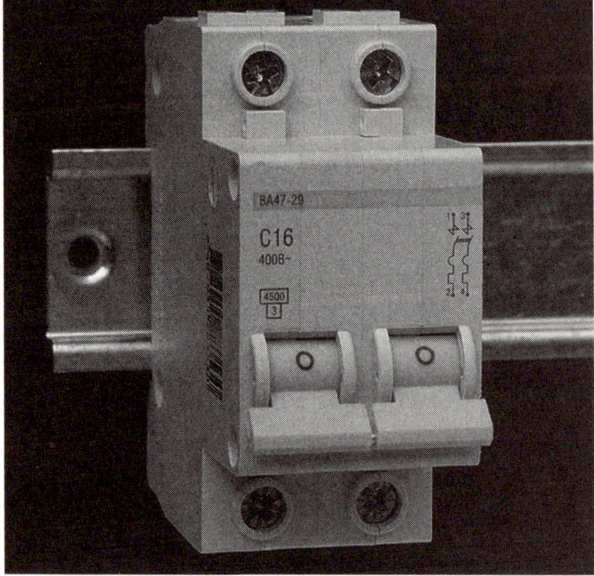

Figure 4.3 Single- and Dual-Pole Circuit Breakers

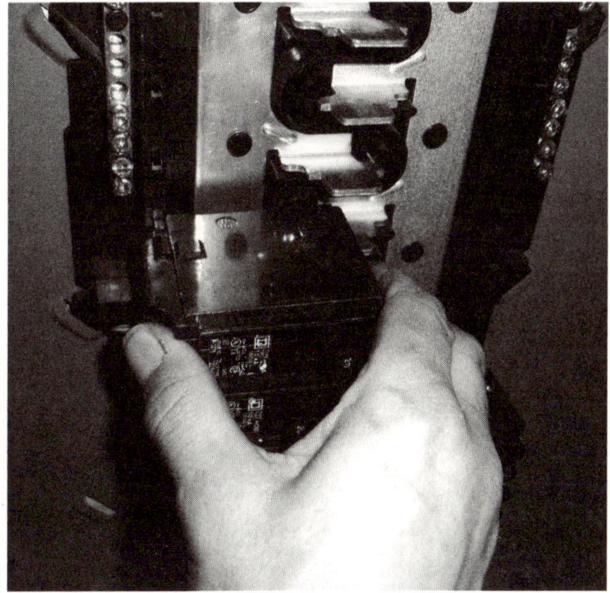

Figure 4.4 Breaker Insertion into a Panel

two breakers can reside in one full-sized slot. While the latter approach allows more circuits in smaller cases, the devices must be appropriately listed.

Unlike in fuses, a screw and a flange connect a hot wire directly to the body of a circuit breaker via a channel. Once this wire is connected, the electrician inserts the breaker end toward the panel's center and snaps the front of the breaker into place, locking it into position (Figure 4.4).

> **Hints and Tips**
> When wiring a circuit breaker, insert the circuit's ground and neutral into the appropriate bus close to the breaker. That way, all circuit wiring bonds to one area in the panel. While NEC does not require this configuration specifically, it does require that grounding potential be equal throughout a grounding system.

▶ Exercising Ground Fault Detection and Protection

Circuit breakers safeguard circuits only in the event of long-term, significant overcurrent conditions. To provide an added layer of protection, the NEC developed regulations for ground fault detection and protection.

In simple terms, a ground fault occurs when the current entering through a hot line differs from that received through the neutral wire. In a properly wired system, the two quantities are the same. The principle is much the same as that of a properly working plumbing system. The amount of water exiting plumbing should match that being carried to the sewer. When more water is exiting the plumbing than entering the sewer, there is either a leak or an excess in the system.

In electrical circuits, current leaks can occur in damp basements, for example. There, the condensation that forms on wiring provides a path to ground that is not through the neutral wire. Even when the current

ALERT!

Because electrons travel at nearly the speed of light, things can go wrong almost instantly. Even 1 millisecond of a 100,000V pulse will destroy every piece of electronics in a residence, and that is far too little time in which any circuit breaker can respond. Circuit breakers are designed to protect property, not life. They simply register the amount of current they are passing and disconnect at appropriate limits. They lack the capacity to identify current's target destination or to compare current entering to current returning. Therefore, they can allow a person to receive a lethal amount of current, which is as little as 10 milliamperes (0.010A). For all these reasons, exercise extreme caution when working with and around circuit breakers.

through such a path is minimal, ground fault circuit interruption (GFCI) breakers trip. GFCI breakers fail to trip when the current is above a standard breaker's rated value.

In contrast to an electrical leak, an electrical excess can occur when the metal frame of an ungrounded appliance contacts a hot conductor that has disconnected in an appliance. The metal chassis becomes charged with energy that has left the circuit but has not yet returned to the breaker. A GFCI breaker will trip; a standard breaker will not.

The nature of GFCI device construction and function requires that wiring procedures for GFCI breakers differ from those for standard models. For example, GFCI breakers have extra terminals to connect

ALERT!

Take extreme care to wire a GFCI breaker correctly. Like receptacles, GFCI wiring is *not* interchangeable. All wires must be installed according to the breaker's markings. While NEC does not formalize such a requirement, it does state that to be listed or marked most devices must be used according to manufacturer instructions and data sheets. For background information, consult NEC Article 100.

to both circuit neutral and panel neutral, whereas standard breakers connect only to the black circuit hot wire directly. The remaining wires connect through the bus.

▶ Studying Arc Faults and Related Protection

An **arc fault** occurs when a path of lowered resistance allows the occasional transfer, or jump, of electrons between hot and neutral. Consider an improperly installed staple, for example, that penetrates insulation and allows electrons to trickle through the puncture. An **arc-fault circuit interrupter (AFCI)** breaker (Figure 4.5) detects such a breach and disconnects the circuit.

Hints and Tips

It is easy to confuse GFCI and AFCI circuit breakers, because they look and wire identically. To distinguish between the two, refer to the breakers' markings.

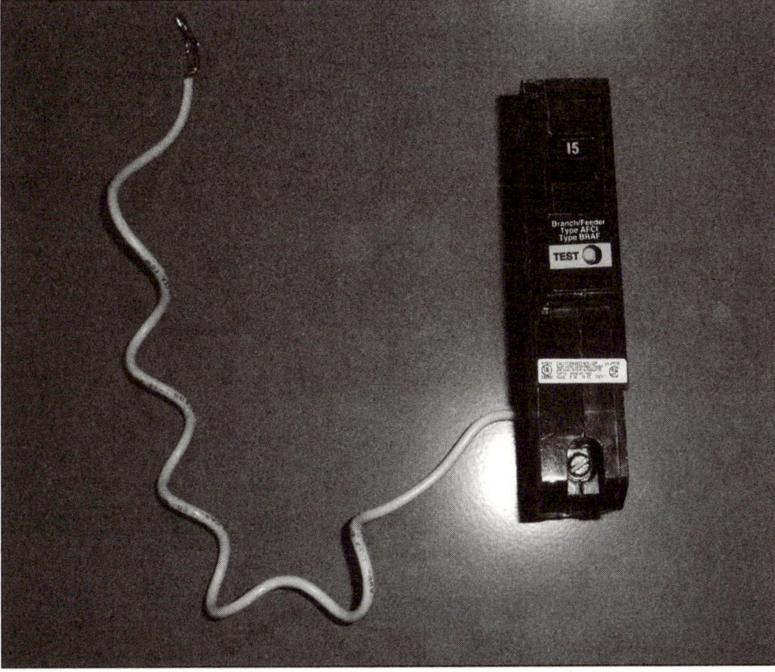

Figure 4.5 AFCI Circuit Breaker and Connections

Did You Know?

AFCI is available in circuit-breaker form only, not receptacle.

▶ Using Smoke and Fire Detection

Whenever electricity is involved, fire is an obvious and serious hazard. When located, installed, and maintained properly, **smoke detectors** deliver an appropriate level of protection from this hazard. These devices are designed to alert homeowners that fire is present or imminent in their structures. When such devices are hardwired to governing alerting systems, authorities, such as fire departments, are seamlessly notified.

A smoke detector must be installed on an undedicated, 15-amp circuit. In addition to **bare-wire ground**, three conductors must be present to provide power as well as signaling functionality. In terms of configuration, the hot wires are spliced together, the alarm (red) wires are spliced together, and the neutrals are spliced together (Figure 4.6). The bare-wire ground passes through the box with no termination(s) unless required by the smoke detector model, some of which have ground lugs. In essence, a smoke detector is wired like a 240V circuit except that (1) bare-wire ground is connected to the detector only when a corresponding plug or pigtail is present, and (2) the red wire carries a signal between all detectors on the circuit, not a double dose of electricity.

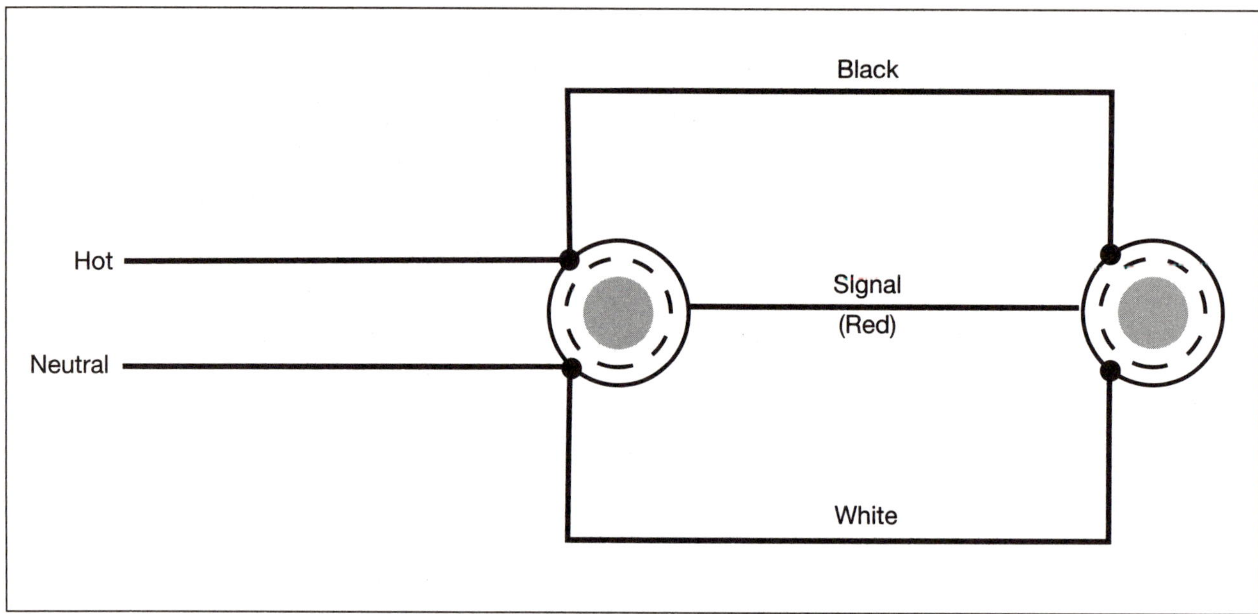

Figure 4.6 Smoke Detector Wiring

The red wire in a smoke-detection circuit travels between detectors so that all detectors sound when one detector does. As a result, these devices alert occupants to a fire anywhere in a residence, which is particularly helpful when those occupants are sleeping. The red wire can also be routed to an electronic interface that can sound an outside siren, send a prerecorded message to 911, or activate other visual or audible alarms. In many new residences, smoke-detection circuits interface with security systems, and many high-end security systems incorporate their own low-voltage smoke-detection circuits.

Did You Know?

The NEC is part of the National Fire Prevention Association (NFPA).

Other types of detectors, like those for carbon monoxide (CO) and radon detection, are wired like smoke detectors, so common sense, manufacturer

Did You Know?

While Building Officials and Code Administrators International, Inc. (BOCA), not NEC, sets smoke-detector requirements, these devices must follow NEC wiring and installation practices. To deliver maximal security, BOCA requires residential smoke detectors in each bedroom, just outside each bedroom, and on each level of the home. For new construction, BOCA requires a 120V AC hard-wired to authorities and interconnected with DC backup. To identify all unique detector stipulations, poll local municipalities.

listing and labeling directions, and local requirements are the guides to installation. Because CO is a by-product of combustion, CO detectors should reside between sources and occupied areas. Typically, CO detectors are placed in a basement near the stairs or at the top of a basement stairway. Massachusetts is one state that requires a CO detector in every residence. The bordering state of Rhode Island does not. As always, consult local fire codes if you are not sure.

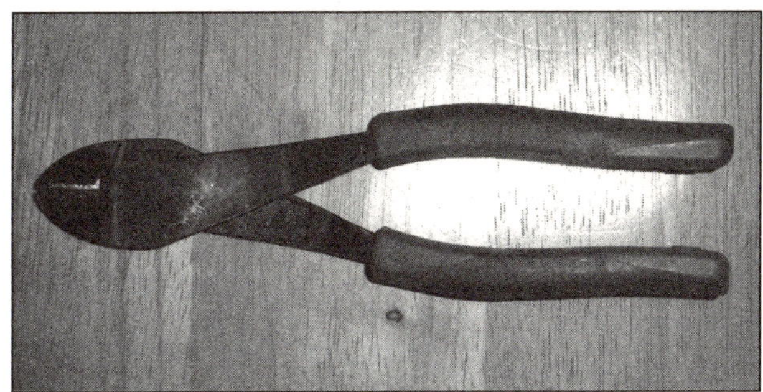

Figure 4.7 Electrician Cutters

▶ Providing Personal Electrical Protection

No discussion of electrical protection is complete without a review of personal safety measures. To illustrate some potential pitfalls, consider the simple task of replacing a circuit breaker on a hot panel. This task requires the electrician to first turn the circuit breaker off and then unscrew the circuit hot wire and pry the breaker out of its enclosing panel. After that, the electrician must install the new breaker, replace the circuit hot wire, and energize the circuit.

Even such a simple task can be filled with risk. Assume, for example, that the breaker panel is in the corner of a wet basement. If the electrician is standing on the wet floor when her screwdriver slips on a corroded screw and accidentally contacts the panel's hot bus, she could easily collapse in front of the box from the shock. In this scenario, as in so many others, one unsafe element does not itself spell disaster. Disaster comes when unsafe acts combine with one another.

Mitigating risk is the responsibility of all electrical workers. Consider the wet basement floor in our example. The electrician could have mitigated that risk by properly inspecting her boots, which should be certified to insulate up to thousands of volts. Unfortunately, our electrician's boots were punctured and provided an electrical channel. The best solution to the corroded screw on the breaker would have been to shut off the master disconnect and make the repair. The next-best choice would have been to cut the wire with a certified pair of cutters or lineman's pliers (Figure 4.7), remove the breaker, and complete the repair. Similarly, electrician's gloves (Table 4.1) are rated by their dielectric value and afford the electrician personal protection when they are worn during electrical work.

All these examples demonstrate that, to be effective, prevention must be proactive, not retroactive. A positive, forward-looking attitude, which includes a respect for electricity, as well as the proper protective equipment, helps ensure that electricians can conduct their work safely, free of significant risk.

Table 4.1 Certified Electrician Gloves

CLASS	TEST AC V	USE AC V	USE DC V	LABEL COLOR
0	2500	500	750	Beige
0	5000	1000	1500	Red
1	10000	7500	11250	White
2	20000	17000	25500	Yellow
3	30000	26500	39750	Green
4	40000	36000	54000	Orange

▶Chapter Review

1. Overall, fuses are easier to work with than circuit breakers.

a. true

b. false

2. What is an ARC fault breaker, and where in a residence is it required?

____

____

____

____

3. Which of the following fuse types are threaded to prevent mismatches?

a. screw-in

b. S type

c. F type

d. barrel

4. ____________ breakers have extra terminals to connect to both circuit neutral and panel neutral, whereas ____________ breakers connect only to the black circuit hot wire directly.

5. GFCI and AFCI circuit breakers look identical.

a. true

b. false

6. Name five residential locations that require GFCI protection.

____

____

____

____

7. Garage-door receptacles require GFCI protection.

a. true

b. false

8. In a residence, a(n) ____________ is an area with a sink.

9. A home with four bedrooms on its second floor with one hall area, one bedroom on its first floor with two hall areas, and a basement, requires how many smoke detectors?

a. two

b. three

c. five

d. nine

10. Smoke detectors are installable on separate circuits.

a. true

b. false

11. ____________ is the organization that sets smoke-detector requirements.

12. The NEC is part of which of the following organizations?
  a. Department of Environmental Conservation (DEC)
  b. Occupational Safety and Heath Association (OSHA)
  c. National Fire Prevention Association (NFPA)
  d. International Electrical Wiring Federation (IEWF)

13. A hallway connecting a bathroom and a bedroom requires arc fault protection.
  a. true
  b. false

14. Current is lethal at which of the following ampere ratings?
  a. 0.0001
  b. 0.01
  c. 1.0
  d. all of the above

15. A non-GFCI-protected wire from a house to a post light must be buried ______________ deep.

16. A water fountain requires GFCI protection.
  a. true
  b. false

17. Why is 14-3 wire required between smoke alarms?

____

____

____

____

18. Placing an electric baseboard heater below a receptacle outlet is sound wiring practice.
  a. true
  b. false

19. The ______________ is the original protective device in household circuitry.

20. All outside fixtures that are prone to moisture should be marked for wet locations.
  a. true
  b. false

21. Identify some safety tips for wiring in a hot panel.

____

____

____

____

22. A bathroom candle receptacle requires GFCI protection.
  a. true
  b. false

23. The resistance between a fuse's two terminals should be ______________ ohms.
  a. 0
  b. 15
  c. 200
  d. 2,000

24. A hot-tub motor requires GFCI protection.
  a. true
  b. false

25. In a fuse, a(n) ____________ is a thin piece of metal that has a low melting point.

26. The motor in a whirlpool tub requires GFCI protection.
  a. true
  b. false

27. How do dry and wet niche lights differ?

____

____

____

____

28. A bathroom fan light requires GFCI protection.
  a. true
  b. false

29. A bedroom fan/light requires arc fault protection.
  a. true
  b. false

30. A GFCI trips in the time period of ____________.

PART

II

At the Office: Commercial Electricity

CHAPTER

5

General Commercial Electrical Requirements

CHAPTER SUMMARY

Commercial electricity represents a departure from residential electricity in a number of ways, although the two approaches are aimed at the same intention: to deliver electricity to a structure. In addition to packaging electrical energy differently, commercial electricity harnesses unique wiring schemes that accommodate the high-demand systems and appliances of commercial environments. Commercial systems and appliances require added sets of conductors with large ampacities, as well as a more robust service entrance; different ratings and types of panels, boxes, and device controls; and a different approach to safety. Commercial transformers, for example, require proper grounding and bonding, and those transformers must be properly connected to loads to ensure safety. When all metallic items are grounded or bonded, power is metered to the installation appropriately, the grounding system provides a closed loop for electrical energy, and commercial feeders and panels are installed properly, commercial applications can carry varying loads without the risk of such hazards as explosions.

Learning Objectives

- Outline the scope of commercial installation
- Compare residential installations with commercial ones
- Define the term *raceway* and its applicability to commercial installations
- Explain the importance of choosing an appropriate raceway and box fill
- Identify the typical branch circuit requirements of commercial installations
- Discuss the purpose of a pull box
- Describe a typical commercial service entrance
- Contrast underground service and overhead service
- Compare different conduit types
- Name applications for flexible and nonflexible conduit
- Identify applications for metallic and nonmetallic conduit

Key Terms

arc blast
bend
bonding
bonding bushing
center tap
coincidental load
continuous load
explosive area
feeder
flexible metal conduit (FMC)
grounding electrode conductor
isolated ground receptacle
noncoincidental load
noncontinuous load
panelboard
primary metering
pull box
raceway
rigid metal conduit (RMC)

Differentiating Commercial Electrical Work

In many ways, residential electrical services and commercial electrical services are like passenger cars and eighteen-wheelers: Both serve the same, generic function with similar methods, but how they do it and in what scope vary. Generally speaking, for example, commercial installations do not use Romex, because Romex cannot easily accommodate many different conductors with different points of connection. Romex's 12- and 14-gauge wiring tends to lack the robustness needed for many commercial applications. Furthermore, it tends to complicate the task of bending when bound with other, thicker conductors. Instead of using Romex, commercial installations feature single conductors through raceways like tubing or conduit. In a sense, commercial electricians make their own Romex by routing conductors through protective metal or plastic tubing, conduit, or raceway (Figure 5.1). The term **raceway**, which refers to any enclosure that routes conductors from place to place, includes conduit of all types.

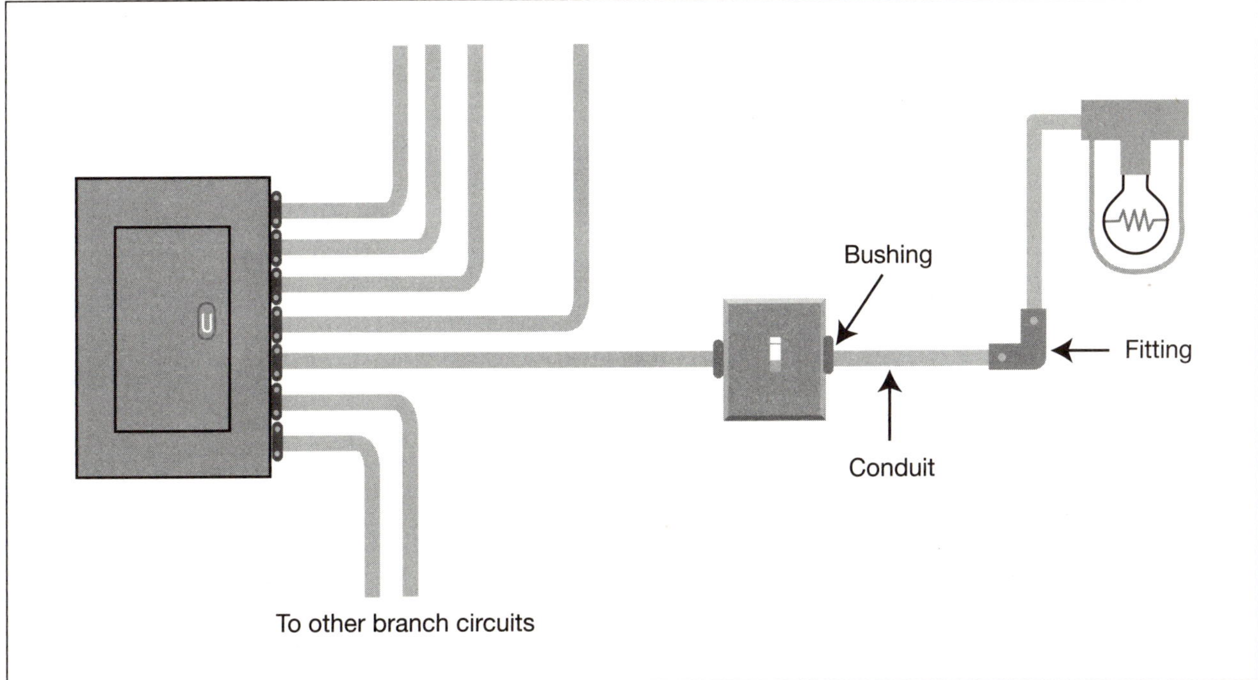

Figure 5.1 Typical Conduit Installation

In addition to using unique wiring, commercial systems use electrical energy that is packaged differently. In most cases, residential wiring is limited to two separate, 120V sources that can be combined to create a 240V source for such high-powered appliances as electric boilers, some furnaces, and most ovens, ranges, and stoves. While commercial environments need 120V sources for office equipment, computers, and lighting, higher voltages (e.g., 480V) and more phases (i.e., similar sources split by time) may be needed for heating, air conditioning, and refrigeration (HCAR) systems and other high-demand appliances.

▶ Identifying Unique Commercial Demands

The high-demand systems and appliances of commercial environments require additional sets of conductors with large ampacities, as well as a more robust service entrance; different ratings and types of panels, boxes, and device controls; and a different approach to safety.

The type of electrical service in a commercial environment depends on the type of service available from the local serving utility as well as the application. Electrical services enter commercial buildings based on local power company guidelines, which may differ from National Electrical Code (NEC) requirements based on regional geographical, climatic, or seismological characteristics. For example, underground conduit must be buried differently in earthquake-prone areas than in low-lying, stable ones. When local guidelines do differ, they take precedence over NEC requirements (NEC Section 90.2).

Once electrical service enters a commercial building, either from the ground or a utility pole, it typically connects to a transformer, which may be pole mounted. Because commercial electrical service is far more robust than its residential counterpart, it may require separate transformers. Recall that transformers are used to step voltages up or down. Given their role, transformers tend to operate with unique rules. Touching a fluorescent light is safe, for example, but touching a long-running fluorescent-light transformer risks a burn.

▶ Outlining the Scope of Transformers

Typically, transformers are available in one of two types: liquid filled and dry (NEC Article 450, Sections 450.21 through 450.28). The liquid in liquid-filled transformers, which is often an oil, serves two important functions. First, it acts as a thermal conductor, removing heat from the transformer's coils and core. Second, it provides insulation. In short, the liquid in a liquid-filled transformer not only dissipates heat, but it also provides a dielectric to keep the hot side of the transformer hot and the cold side cold. Cooling fins aid in transferring heat from the transformer's coils or core by liberating heat into the ambient air.

Like liquid-filled transformers, dry transformers have cooling fins. Unlike their liquid-filled counterparts, however, dry transformers rely on air circulation and an air gap, not a liquid, to provide cooling and insulation. When weight is a consideration, dry transformers are preferred to liquid-filled ones because they are lighter. However, dry transformers must be installed in open areas to facilitate cooling-fin ventilation. In addition, they must be at least 12 inches from combustible items (NEC Article 450.21). Closed areas without

Did You Know?

Askarel, a mixture of polychlorinated biphenyls (PCBs), was once a common dielectric choice for liquid-filled transformers. In fact, some old transformers still use this liquid. In 1976, however, the Environmental Protection Agency (EPA) deemed it a hazardous substance, thereby restricting its use in new transformers.

ALERT!

Because oil is a flammable liquid, improper installation of a liquid-filled transformer poses a safety hazard. Leaking liquid can cause a catastrophic fire either due to its inherent acidity or its contact with electricity. In most cases, improper installation manifests as insufficient ventilation or cooling, leading to viscosity changes and premature wear of seals and gaskets. A proper installation can degrade to improper status when it has (a) been superseded in code or (b) deteriorated to a point at which it is no longer proper, such as when a ventilation fan in a transformer vault or casing burns out. To mitigate the risk of liquid-filled transformers, inspect installations whenever possible.

Hints and Tips
Oil-filled transformers installed indoors must be constructed in concrete areas, because their hazardous substances may leak. For specific requirements, see NEC Sections 450 and 505.

Did You Know?

The most common areas for oil-filled transformers are large commercial buildings like shopping malls. The transformers in oil fields have windings immersed in highly refined mineral oil, silicone fluid, or an edible seed, oil-based dielectric coolant. Some oil-field transformers use nontoxic liquids.

proper airflow are not options. Mechanical or outdoor areas are preferable, although outside transformers require protection from the elements. When local guidelines dictate, some transformers are enclosed in underground vaults. Underground vaults are typically used to maintain architectural flow or to save space. Any secondary transformer that is outside typically enters a building through a large cable assembly.

When dry transformers must be placed close to combustible items, fire-resistant material is required as a buffer (NEC Article 450.21). Furthermore, fire-suppression systems should be installed. In general, any such fire-suppression or fire-extinguishing system must be able to run for one hour or more per fire codes established by the National Fire Protection Association (NFPA). For full, governing regulations, defer to local authorities.

Did You Know?

Dry transformers are used most commonly today because they require less maintenance and pose less of a hazard than their liquid-filled counterparts.

ALERT!

Much like transformers in fluorescent light fixtures, dry transformers can get hot, approaching the rated limits of the associated conductors. Sunlight can increase overheating, so proper transformer locale and proper ventilation are critical.

Hints and Tips

When repairing any transformer, always check with local authorities for the wiring requirements that take precedence. In general, however, NEC Articles 250 and 450 govern transformer installation. NEC Article 450 describes the classification of transformers by voltage level. For transformers over 600V, see NEC Table 450.3(A); for transformers of 600V or less, see NEC Table 450.3(B).

Grounding and Bonding Transformers

The proper grounding of electrical current dictates a **grounding electrode conductor** that is as short as possible to ensure the fault current has the least resistance and therefore the fastest path to Earth ground. When electrical equipment is never to be energized, its outer metal casings are bonded.

Most power companies require one or more ground rods at a transformer. The NEC requires a

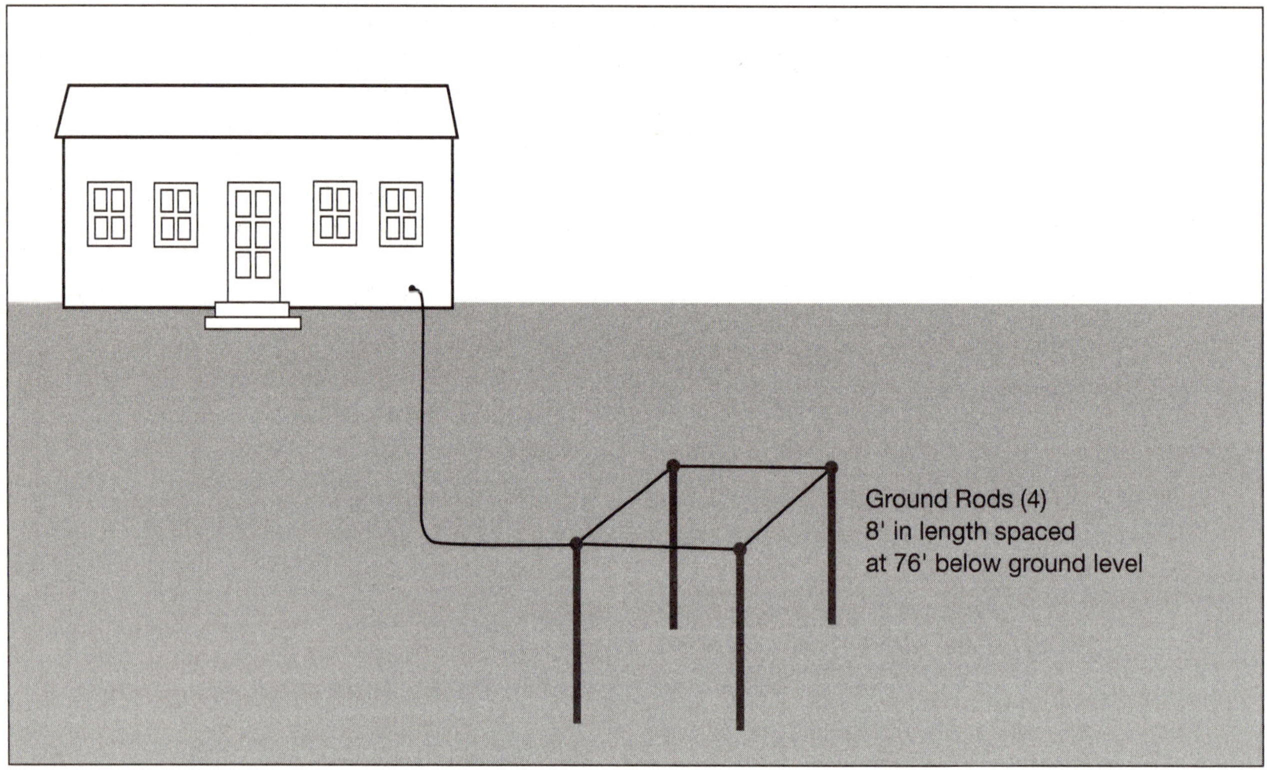

Figure 5.2 Grounding Grid

grounding electrode conductor from service equipment to a metal water piping system within the first 5 feet of the point of entrance or to the metal structure of the building, whichever is closer (NEC 250.30). Every electrical installation differs, however. While many small installations work well with single ground rods, large installations may require grounding grids (Figure 5.2). A grounding grid is installed in a 3-foot-deep hole into which several ground rods are hammered and bonded together before **bonding** to the equipment being protected. A transformer combined with a high-voltage switch and a distribution array creates a transformer substation (Figure 5.3), the purpose of which is to step down voltages, increasing available current, and to provide routing control for the resulting power to be distributed.

Hints and Tips

In any grounding electrode system, ensure that all equipment and structural grounds are connected at only one point. Doing so prevents stray voltages in the event of lightning strike.

The transformers used in commercial applications change the voltages entering buildings. Some buildings, for example, require step-down transformers to achieve voltage decreases from 480Y/277 or 208Y/120. Outside a building, some heavy machinery requires 480Y/277 and some lighting requires 277V. Inside a building, 208Y/120 may be needed to power 208V water heaters, while 120V may be needed for photocopiers. To afford lighting and appliance **panelboards**

Figure 5.3 Transformer Substation

Hints and Tips
Remember that high voltages require less amperage to transmit the same power.

overcurrent protection on their supply sides as well as maintenance and disconnecting means, fused disconnects can be required locally before and after transformers.

Connecting Transformers to Loads

Transformers connect to loads in a number of schemes. In the first, a single-phase system that is identical to that used in residential applications, a neutral supplies two 120V phases (Figure 5.4). This neutral, called a **center tap**, must be grounded (NEC Article 250.20(B)(1).

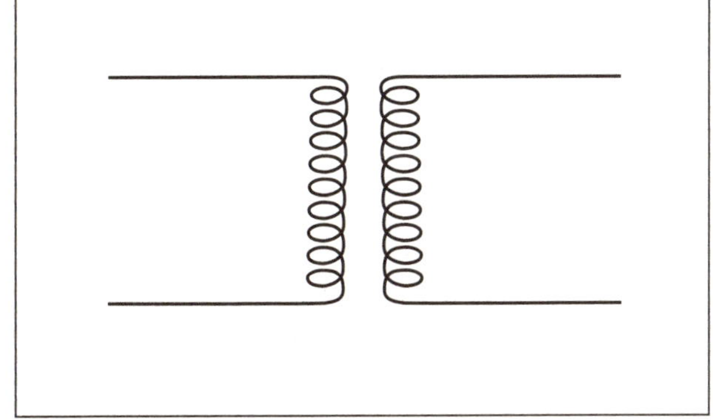

Figure 5.4 Single-Phase System

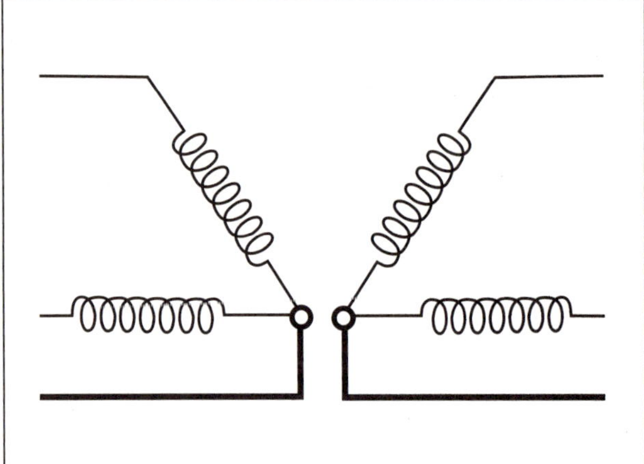

Figure 5.5 Open Delta System

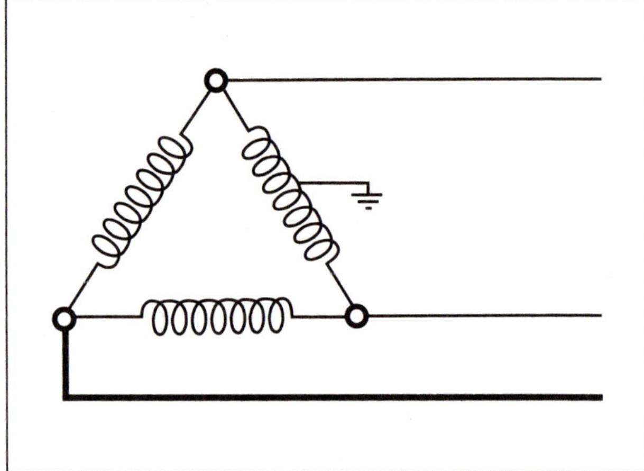

Figure 5.6 Four-Wired Delta System

The next scheme, called an open delta system, allows for three-phase power using just two transformers (Figure 5.5). The four-wired delta system, in contrast, uses three transformers to provide three-phase electricity (Figure 5.6). In this scheme, one of the transformer's **center taps** functions as the ground wire/neutral terminal for a single-phase adjunct system.

The three-wired delta system, used in commercial applications as well as industrial ones, provides only three-phase power with no provision for single-phase power output (Figure 5.7). In this scheme, any of the three legs carries 480V. Any two legs should register 480V compared to the third leg, which is grounded. One ungrounded leg is measured at 480V, while the two ungrounded legs are separated by 277V. No overcurrent devices should be installed in the grounded phase.

The final scheme for connecting a transformer to a load, the three-phase four-wire wye system (Figure 5.8), is commonly used in commercial installations

ALERT!

The three-wired delta system is a great way to power three-phase equipment. When running circuits into the panel box, however, check each phase to ground to avoid installing a 277V piece of equipment. This schema will destroy any such equipment.

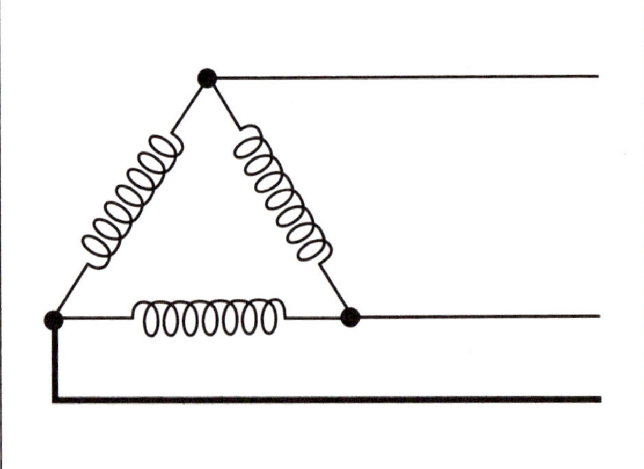

Figure 5.7 Three-Wired Delta System

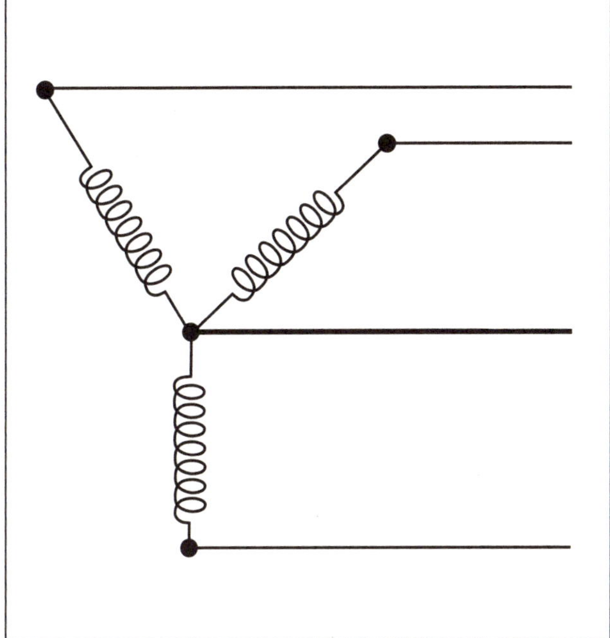

Figure 5.8 Three-Phase Four-Wire Wye System

because it is the most flexible. In addition to providing three-phase power, the three-phase four-wire wye system allows connection of any single phase as a 120V source. While this system offers 120/208 and 277/480 voltages, the turns ratio determines specific output voltages. Transformer connections are similar regardless of output.

► Metering Power to Installations

Like wiring the service entrance, metering power to an installation is an important aspect of commercial wiring. While electricians rarely interface with metering directly, knowledge of the metering types typically used in commercial installations is beneficial.

Simply put, power usage is measured before or after a transformer. Generally, a meter has two transformers: (1) a potential transformer that senses the voltage being delivered and (2) a current transformer that, through a coil, converts current to voltage that can be measured. The meter calculates a power value by multiplying voltage and current.

Meters are classified two ways: (1) high voltage and (2) low voltage. In single-use commercial installations, which have single owners or tenants, it is sometimes easiest to monitor the high-voltage side, because one customer receives all voltage. When, in contrast, multiple customers receive voltage, that voltage is typically subdivided at the low-voltage side. The output of each subdivision must be metered separately.

A commercial installation with many rows of meters demonstrates low-voltage metering, often called secondary metering. An installation using high-voltage metering, often called **primary metering**, typically has only one meter. For the most part, power in primary metering is the same as in secondary metering.

Hints and Tips
The current on the high-voltage side of a transformer is significantly less than on the load side because, per Ohm's law, power equals voltage times current. Power is equal on both sides. If voltage is lower on one side, current must be higher to maintain this equality.

Hints and Tips
The breakout from service equipment to a tenant is that tenant's **feeder**. Each feeder, which links service entrance equipment and a panelboard, manages a particular subinstallation.

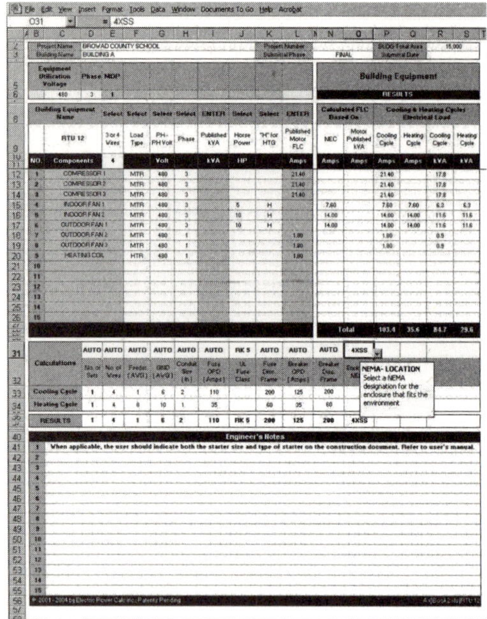

Figure 5.9 Sample System Configuration

Depending on the size, scope and application of the installation being considered, the resultant design may be too complex to be committed to memory, or to scraps of paper. Because of this, there exist many different software packages designed to assist the electrician in coming up with the most effective solution to these complex installation requirements. A screenshot of one such software package is shown in Figure 5.9. The electrician simply enters in the requirements for branch and feeder circuitry, and the software computes the necessary parameters.

It is not uncommon for the sum of feeder circuits' ampacities to exceed a meter's ampacity rating. This is because circuits and load types are classified not only by application but by percent usage. For example, an installation may allocate energy to a cooling system and a heating system, but these two systems are unlikely to be active at once. Additionally, motors at startup require more current than they do at running speed. As a result, current draw increases at startup but then gives way to a lower constant average draw once momentum is achieved through rotation and counter electromotive force (EMF).

> **Did You Know?**
>
> Electricians must be able to disconnect commercial service to a building using no more than "six throws of the hand," meaning they must throw no more than six switches to disconnect a building's power (NEC Article 230.71).

▶ Grounding Electrical Service

In a commercial installation, all metallic piping, ducting, siding, and other metallic items that could contact electrical energy should be grounded or bonded (NEC Article 250) (Figure 5.10). As mentioned previously, grounding or bonding limits the potential to ground by providing a means to sense overcurrent and subsequently a means to disconnect power from a circuit in the event of a problem (NEC Article 250 Part 2).

Given that a good grounding system provides a closed loop for electrical energy, all grounding contacts should converge at one point to prevent any opportunity for different ground connections to exhibit different voltages. Per Kirchoff's voltage and current laws, single convergence points prevent this from happening.

Raceways entering and leaving main service equipment must be bonded to that equipment through **bonding bushing** (NEC Article 250.92). A bonding bushing, which is a pressure-fit device used to guarantee good contact between metallic surfaces, ensures

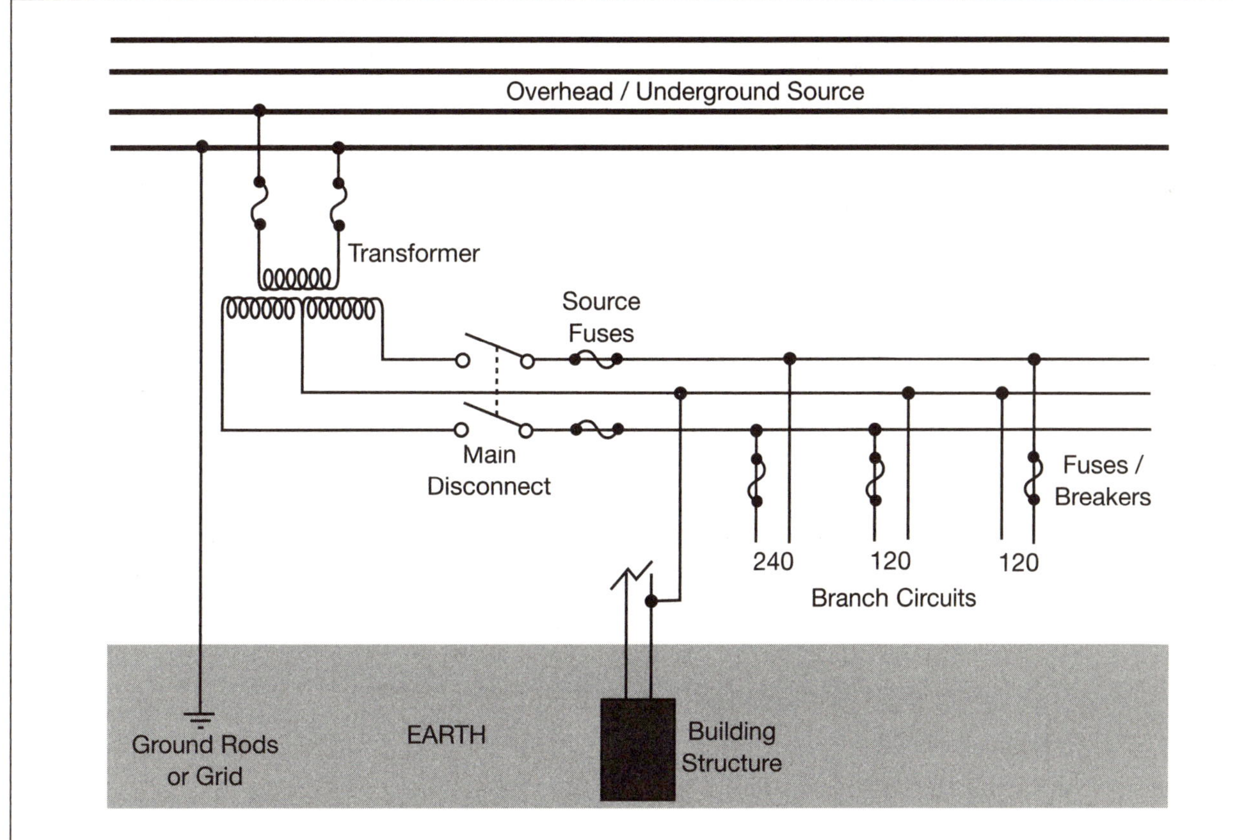

Figure 5.10 Typical Commercial Grounding Connections

that the main service equipment and the raceways attached to it are at the same voltage level or potential. In essence, bonding bushing ensures that two metallic structures lack resistance.

Service equipment houses a neutral bus and a grounding bus that, in addition to being tied together, must be bonded to the earth. Connected to this grounding bus is a conductor, called a grounding electrode conductor, which connects the grounding bus to the water service entrance metal plumbing in the building on both sides of the water meter and within a specific distance from the building's entrance as determined by local codes. Any hot water piping must be electrically bonded to cold water piping to ensure both

Hints and Tips

Kirchoff's voltage law simply states that all voltages must be accounted for. In other words, when voltage is where it should not be, it must be balanced by a lack of voltage elsewhere.

ALERT!

Explosions occur when a flammable gas meets an electrical spark.

Did You Know?

Connecting a ground to a piece of plumbing does *not* constitute grounding an entire plumbing system (NEC 250.104[A]).

that interior plumbing infrastructure is bonded to the entire electrical service entrance and that shocks from hot wire contacting metal plumbing are prevented (NEC Article 250).

When establishing a bonding or grounding system, a number of other NEC requirements apply. For example, although a metal underground gas pipe should be grounded, it cannot be the primary grounding conductor or electrode (NEC 250.104[A]). Per NEC 250.64, grounding electrode conductors may not be spliced unless welding or compression-type fittings prevent removal.

The metal frame of any building must be grounded or bonded to the grounding electrode (NEC 250.50), and grounding jumpers are required across all insulating joints in a piping system (NEC 250.104[A]).

In other NEC requirements, the grounding electrode must be connected to the supply side of the service disconnect (NEC 250), as well as accessible, unless it is buried or encased in concrete (NEC 250.50–52). In the latter case, the bonding point of the grounding conductor must be accessible (NEC 250.142[A] and [B]). Finally, grounding in bonding conductors must be of sufficient ampacity to route any stray or arc currents from hazardous situations.

Hints and Tips

A grounding conductor must be at least as robust as the largest current-carrying conductor in a circuit. NEC Table 250.66 specifies grounding conductor size based on ampacity requirements.

Unlike in many residential installations, ground fault protection in many commercial applications encompasses entire circuits rather than individual receptacles. (For more on ground fault protection, see Chapter 7.) In most commercial applications, depending on usage, ground fault protection is required in bathrooms, kitchens, and outdoors (NEC 210.8[B]).

▶ Categorizing Commercial Feeder Requirements

NEC Article 220 Part 3 details commercial feeder requirements. One of these is overcurrent protection, which is discussed in NEC 215. While in general, conductors must be protected based on their individual ampacities, when a conductor's ampacity fails to match a standard overcurrent protector rating the next-higher-rated device may be used. For example, when a conductor's ampacity should be 140 amperes, a 150-amp protective device may be used because 140-amp protective devices do not exist (NEC 240.4(B)).

A second requirement for commercial feeders involves wire selection, which amounts to two choices: (1) copper based and (2) aluminum based. Overall, aluminum is less expensive but copper is a better conductor. To carry a given amount of current, larger diameter (smaller gauge or AWG) aluminum wire is required, because it requires more cross-sectional area to conduct the same amount of current as copper wire.

Connectivity is often an issue with aluminum wiring for a number of reasons. First, because it has a larger temperature coefficient than copper, aluminum tends to contract and expand more than copper wire with temperature changes, and therefore, requires more

ALERT!

A catalyst like water fuels the aluminum/copper chemical reaction and amplifies the potential hazard caused by corrosion and different thermal expansion rates. This is why rated devices must be used with aluminum wire and why aluminum wire must be treated before it can be terminated. Devices and terminations are marked "CU" for copper, "AL" for aluminum, and "AL/CU" for both.

robust crimp connectors and more attention to detail from the electrician. Second, on exposure to air, aluminum starts to tarnish or oxidize, jeopardizing sound electrical connections. Electricians must use chemistry to prepare aluminum conductors for use, which entails applying a solution to the bare wire, allowing it to work, and brushing it off.

As with copper wire, a torque wrench is needed to properly secure an aluminum conductor to its mount, and special connectors and procedures are needed when connecting aluminum and copper wires. Generally, aluminum wire is used only with the very large, large-gauge conductors used most often near service entrances, because aluminum and copper can react chemically, causing corrosion, impeding conductivity, and changing the wires' shapes and sizes.

Temperature is a third general requirement of commercial feeders. A device's temperature rating tends to govern the device's ampacity based on the ambient temperature throughout a given period, and a similar paradigm applies to conductors. Generally, conductor size is based on the temperature rating of a circuit's terminations (NEC 110.14), which are rated for a certain temperature (e.g., 140°C or 167°C).

A fourth and final commercial feeder requirement is feeder ampacity. First, a feeder's current-carrying capacity must be greater than the total coincidental current as calculated for individual branch circuits according to demand and duty cycle (NEC Article 220 Part 2). Second, when two currents are not coincidental, the larger of the two currents should be used in calculations (NEC 220.60).

Following is the process for calculating feeder ampacity.

1. Using NEC Table 310.13, determine the conductor type.
2. Identify the number and sizes of conductors to be installed.
3. Determine the highest ambient temperature of the installation.
4. Conduct continuous and noncontinuous ampacity requirements ("circuit loading") (NEC Article 220).
5. Calculate overcurrent protective device (OCPD) ratings for each feeder and/or branch circuit per NEC Section 240 (noncontinuous + continuous × 1.25).
6. Determine conductor size based on temperature and ampacity requirements per NEC Table 310.16.

Working with Commercial Panelboards

Unlike residential installations, which typically feature single panelboards, commercial installations allot single panelboards for each tenant or office area. From the main service equipment, these panelboards are routed individual feeders that terminate in a main panelboard. The protection rating of the overall commercial feeder circuit must meet or exceed the sum-total of the

Did You Know?

Panelboards ranging from 100 to 600-plus amperes are common. A panelboard's ampacity depends on, and must match or exceed the ampacity of, the main bus supplying it (NEC 408.30). For a panelboard to remain listed, it must have matching components (NEC 230.8). In polyphase panelboards, each phase has its own bus, and bus arrangement must be stated and obvious.

calculated loads fed from the circuit (NEC 230.79–80). When connecting multiple panelboards to an individual feeder, either the first panelboard can serve as a junction box for a second panelboard or a separate junction box can be installed to feed both panelboards.

To ensure proper maintenance and inspection, a panelboard and/or service entrance must be afforded adequate access (NEC 230.72[C]), as well as an adequate means of escape (NEC 110.26). Specifically, panelboard or service-entrance working space must be at least $6\frac{1}{2}$ feet high, the access area must be more than 30 inches wide, and the surrounding space must be uncluttered by such obstacles as heating, ventilation, and air conditioning (HVAC) ductwork or shelving. In addition to meeting dimensional requirements, a panelboard or service entrance space must be illuminated effectively (NEC 110.26[D]), and that illumination must be controlled manually (NEC 210.70[A][1]).

The complexity of commercial installations renders proper circuit labeling paramount. A diagram that identifies these circuits, called a circuit directory (Figure 5.11), is an important part of overall system documentation, because it serves as a road map for electricians completing future work on the installation.

LP-01 Electric Services, Inc (555) 555-5555	225A MLO 480Y/277V 3PH 4W Room 107 FF: FHDP Electrical Rm 016
LIGHTING ROOM 102 01 15A 1P-1 © #01 A 2720VA	LIGHTING OUTSIDE VESTIBUL 112, 113 15A 1P-1 © #02 A 248VA 02
LIGHTING OUTSIDE VESTIBUL 110, 111 03 15A 1P-1 © #03 B 448VA	LIGHTING ROOM 103 15A 1P-1 © #04 B 2720VA 04
LIGHTING ROOM 104, 105 05 15A 1P-1 © #05 C 2325VA	EXIT SIGNS & EM. LIGHTS ROOM 103 15A 1P-1 © #06 C 77VA 06
EXIT SIGNS & EM. LIGHTS ROOM 101, 105 07 15A 1P-1 © #07 A 142VA	LIGHTING ROOM 106 15A 1P-1 © #08 A 1770VA 08
EXIT SIGNS & EM. LIGHTS ROOM 102 09 15A 1P-1 © #09 B 104VA	EXIT SIGNS & EM. LIGHTS ROOM 104, 106, 107 15A 1P-1 © #10 B 120VA 10
LIGHTING OUTSIDE VESTIBUL 108, 109 11 15A 1P-1 © #11 C 448VA	LIGHTING ROOM 107 15A 1P-1 © #12 C 1770VA 12

Figure 5.11 Sample Circuit Directory

Did You Know?

No hot conductors can be within reach of a person standing in front of a panelboard (NEC 230.62), because hot conductors and grounded conductors look the same to the uninitiated.

Hints and Tips
Place emergency lighting at the panelboard so that, in the event of a power outage, the panel will be accessible and visible.

Hints and Tips
To calculate load values for noncoincidental loads, use as a continuous load value the highest ampacity requirement from the list of noncoincidental loads. Then, multiply that value by 125%.

▶ Identifying Characteristic Commercial Loads

Commercial applications carry varying loads. As its name suggests, a **continuous load**, for example, runs throughout normal commercial business hours to power things like lighting, electrical outlets, ventilating equipment, and telephone and data systems. A **noncontinuous load**, in contrast, is energized only intermittently throughout the business day.

Hints and Tips
When working with branch circuits, consult NEC Article 310 for details on conductors, trade names, ampacity tables, derating factors, and insulation types.

Another type of load, the **coincidental load**, uses energy at the same time as other loads. In a tanning salon, for example, tanning beds and the ventilation system are coincidental loads, because the ventilation system is only on when the tanning bed is on. **Noncoincidental loads** generally do not use energy at the same time. Heating systems and air-conditioning systems are examples of noncoincidental loads. Air conditioning is used in the spring and summer, while heaters are used in the fall and winter.

Hints and Tips
To calculate a circuit's required ampacity, multiply a continuous load by 125%.

▶ Working with Branch Circuit Wiring

Branch circuit wiring leaves a breaker assembly and travels out of the panelboard through conduit bonded to the panelboard's chassis. Whereas $\frac{1}{2}$-inch conduit used to be the standard, $\frac{3}{4}$-inch conduit may now be installed to accommodate wire expansion (NEC Chapter 9 Table 8). For example, $\frac{1}{2}$-inch conduit accommodates just nine number 12 THHN conductors, whereas $\frac{3}{4}$-inch conduit accommodates 16.

For general wiring, electrical metallic tubing (EMT) is the most common conduit type. Compared to its residential counterpart, Romex, EMT can sometimes be difficult to route through a building. Conductors pulled through conduit create friction, which, over long distances or **bends**, complicates the pulling process. When conduit stretches for long distances or negotiates several bends, **pull boxes** are required to

Hints and Tips
When pulling conduit, use stranded wire instead of solid wire, because stranded wire is easier to bend.

Hints and Tips
FMT is usually used in areas permitted by NEC Article 360.10.

allow access inside conduit (NEC 314). Pull boxes are suggested by good practice for all bends exceeding 360°. While 360° is the guideline bend between pull boxes, experienced electricians often self-impose a maximum of 270° for their installations.

The tendency of equipment to vibrate must be considered when establishing connectivity to a piece of mechanical equipment. Some form of **flexible metal conduit (FMC)** is required (NEC 348.12). Greenfield, a common flexible metal tubing (FMT), is corrugated flexible tubing that can be connected to EMT with fittings approved by EMT and FMT manufacturers (NEC Article 360).

FMC is inappropriate in areas that are concrete or subject to physical damage or moisture, because it cannot take the stress and pressure of being embedded in a solid material like cement. Locations requiring FMC, such as machine shops, must not be classified as hazardous (NEC Article 500.3[B]). See NEC Article 500.3B for proper ways to use FMC with corrosives, in vapor-prone areas, and when proximity to a mechanical piece of equipment risks physically damaging the conduit.

While FMT is inappropriate in moist areas or areas subject to conduit damage, it can be used to safely connect junction boxes and lighting as long as fill requirements are honored. FMT can serve as equipment ground, but when possible, try to get a ground wire from ground sourced directly to the equipment being connected.

Did You Know?

Self-manufactured adapters are disallowed, because they void the UL listings of EMT and FMT manufacturers. Connections and terminations must be listed.

Job Connection

While installing a subpanel as part of an office building's renovation, Marcus and Tatiana were asked to run EMT from a basement to a third floor. While trying to install a nonflexible, 2-inch piece of EMT into a wall well, the pair encountered difficulty trying to run a 90° bend into a confined space. To solve the problem, the electricians hooked FMT from the end of one piece of EMT and then ran the FMT at a 90° bend into the confined space. Much to the customer's satisfaction, Marcus and Tatiana were able to complete the job quickly and easily.

▶ Working with Polyvinyl Chloride (PVC) Conduit

Polyvinyl chloride (PVC), the piping used in many plumbing applications, is disallowed in areas susceptible to physical damage or high temperatures due to its brittleness and thin walls, as well as its low melting temperature (NEC 362.10). It is allowed, however, in places like above drop ceilings, behind drywall, and into block or brick walls. Where there may be high corrosion possibility, such as dairy farms, underground services, and concrete plants, a different type of conduit may be needed.

Because ENT is a nonconductive material, a separate equipment grounding conductor must be installed (NEC Article 362.60). In areas of high traffic or damage exposure, a "Schedule 80" must be run with PVC. A Schedule 80 coupling or box, in contrast to a Schedule 40 piece, is thicker, harder, and can sustain more stress before damage occurs. Schedule 40 and Schedule 80 also differ in inside diameter; for a given trade size, Schedule 40 has a larger inside diameter than Schedule 80 (Figure 5.12). ENT couplings come in one style; no interior or exterior coupling or box connector is made.

When PVC runs under concrete at least 4 inches thick, it is considered outside the structure and qualifies as Schedule 40 as soon as it exits the ground. Where it is exposed, however, PVC must be classified as Schedule 80 (NEC 362.120). PVC markings must be visible every 10 feet and visible for inspection (NEC 362.120).

> **Hints and Tips**
> When installing ENT, support it at closer intervals than EMT (NEC 362.30). Strap it every 5 feet, because at the suggested support interval of 10 feet it sags in the middle. Avoid adding wire to ENT, because wire only makes the bow more acute and increases the friction of any conductors being pulled through. Secure nonmetallic conduit using one- or two-hole straps made of PVC or metal. Most electricians will use galvanized one-hole straps marked for rigid conduit, because their outside diameter is about the same as ENT or PVC. The number of bend degrees allowed with ENT are the same as for EMT (NEC Article 362.26), but when bends are 360° consider installing a larger piece of conduit to ease the pulling through of conductors.

> **Hints and Tips**
> Keep Schedule 40 and Schedule 80 differences in mind when calculating wire fill requirements.

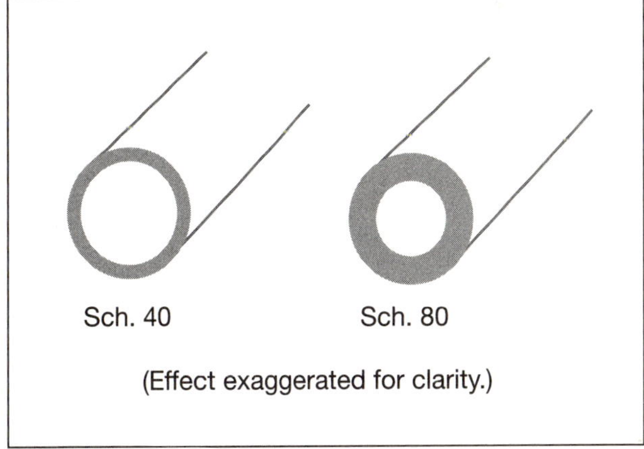

Figure 5.12 Schedule 40 and Schedule 80 PVC Conduit

▶ Installing Weatherproofing

Unlike FMT, both EMT and ENT can be installed in moist locations. Because it is galvanized, EMT is impervious to corrosion. Outdoors, EMT can be painted and installed with compression fittings that allow no water to enter. Outside junction and device boxes must be waterproof and threaded and, like conduit, require weatherproofing and full protection from the elements, which includes ultraviolet light and insects as well as snow, wind, and rain (NEC 314.15 and 110.20). Such boxes vary from $1\frac{1}{2}$ to 3 inches deep; most come in single-, two-, or three-gang metal with screw-in blank plugs.

ALERT!

Water allowed into EMT can prove hazardous to equipment and the public. Any time electricity and water are in close proximity, a hazard exists, because water is an excellent conductor.

Hints and Tips
Fasteners supporting boxes should not serve as ground screws, because the fasteners loosen over time from expansion, contraction, or building settling.

Hints and Tips
When choosing junction and device boxes, review building requirements and prints. Often, it is advantageous to use one manufacturer's products for a job.

PVC is available in the same depths and configurations but lacks a knockout screw-in plug and is inappropriate when explosion-proof conduit is needed. It is brittle and melts at low temperatures. ENT is preferable over EMT in many applications because it is nonconductive, air- and water-tight when sealed properly, and effectively immune to most corrosive agents. When fitted properly, ENT presents a raceway that is sealed from the elements. The boxes installed with ENT are usually sealed with a plastic-style material. These boxes repel sunlight, prevent corrosion from the elements, and repel insects. Installations in marine areas tend to use ENT, because most of the conduit and junction boxes are nonmetallic and therefore exempt from the corrosive effects of sea salt.

Hints and Tips
To prevent fumes from entering the raceway when EMT runs outside a paint booth, install the EMT at least 5 feet from doors and 10 feet from curtains. Coordinate with the appropriate personnel to ensure that, if the fire-suppression system trips, all power to the paint booth will shut off automatically.

Did You Know?

A redundant ground is designed to eliminate an **arc flash**.

Hints and Tips

In certain geographic areas, strapping should occur at intervals shorter than NEC mandates. For details, refer to local requirements. In general, though, as diameter increases, strap more frequently.

▶ Navigating Explosive Areas

Rigid metal conduit (RMC), which varies from $\frac{1}{2}$ to 4 inches in diameter, is designed to seal off an electrical supply from any explosive or combustible entity. Its common applications, therefore, are gas stations, chemical storage areas, and paint booths.

To keep gas fumes from exiting or entering a piece of conduit, panel, or junction box in such hazardous, **explosive areas**, both ends of RMC must be filled with a fireproof sealant. For similar reasons, junction boxes must be cast metal with no exposed holes, which could admit oxygen, and they cannot be mounted to structures except by using flanges or tabs (NEC 314.11). Instead, RMC is mounted to a structure within 6 inches of the box, which effectively mounts the box to the RMC, which, in turn, mounts to the structure. Raceways must be supported starting 3 feet from the box and every 10 feet to the next junction box. Pipe length determines the suggested or required support interval (NEC Table 344.30[B][2]). When conduit must run between two points (from a building to a gasoline pump), for example, thick RMC with coarse threads on both ends is needed.

As mentioned previously, EMT is noncorrosive. As such, it is designed primarily for indoor use where there is little chance of corrosion or damage and it cannot be used outdoors. NEC Article 358.30 Sections A, B, and C address proper EMT installation.

At $\frac{1}{2}$ to 2 inches, EMT must be supported at least 3 feet from the box (NEC 300.30[A]) because the box connector and box fastener are considered part of the support system. Using a threaded fastener, securely fasten grounds into the metal framing of the junction box or into a box connector. The one- or two-hole metal straps of fasteners must be 3 feet from the box

Hints and Tips

For all procedures, check with local authorities for governing requirements. Local regulations supersede NEC.

ALERT!

Failing to bond and ground all metal areas when running conduit risks arc flashes.

ALERT!

The all-oxygen atmospheres of many medical facilities make patient breathing easier but pose a significant electrical hazard. Should an arc flash occur, the resulting fire will grow very quickly and become unmanageable.

Hints and Tips

Receptacles for isolated grounds typically have green triangles on the lower right surface of the receptacle as viewed from the front. Sometimes, these green triangles appear on an orange background.

and then every 10 feet from strap to strap (NEC 314.23[H][2]).

Metal conduit 5 and 6 inches in diameter is cumbersome in any length. When the conduit runs through a wall, the wall may act as a support, but the raceway must be one piece. While some electricians use the conduit as a grounding conductor, better practice is to run a separate ground inside the raceway (NEC 250.118). Circuits designated for computer use should have an isolated ground (NEC 110.27[A]). **Isolated gound receptacles** designed for use in an isolated ground carry small triangles on their faces (Figure 5.13).

Some manufacturers now color-code conduit based on intended use so that emergency personnel and technicians can immediately determine the hazard level of conduit and its contents. Orange conduit, for example, is reserved for fiber-optic raceway that does not carry 120V or 277V power. Blue conduit is data-communication cabling that carries no line current or voltage. Gray conduit carries power at 120V or 277V, whereas green conduit is used for such healthcare applications as hospitals and dental and doctors' offices. Yellow conduit, found in most industrial applications, is reserved for high voltages. Although NEC does not yet require color-coded conduit, it likely will in time.

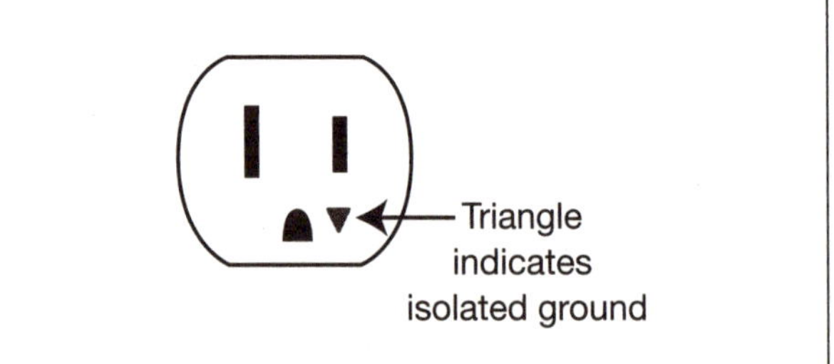

Figure 5.13 Isolated Ground Receptacle

▶ Wiring Healthcare Facilities

Healthcare facilities, one special wiring application, require hospital-grade metallic conduit (AC cable) that is easily identifiable by its color (a light green) and a

Hints and Tips

Because open garage doors often conceal exit signs, do *not* use them as emergency exits.

Hints and Tips
When developing the circuitry and raceways supporting a healthcare facility, consider the number of emergency bulbs and watts per bulb, as well as protective elements, and the option of using low-energy LED lighting.

Did You Know?

Emergency lighting need not be as bright as standard lighting, because it is only used to guide occupants out of a building.

redundant ground, which is a ground on the wiring's inside plus a ground for the outer casing. In addition to these items, healthcare facilities require GFCI protection in waiting rooms, operating rooms, bathrooms, and food-preparation areas (NEC Article 517.20[A]). Like other commercial areas, medical facilities also require emergency lighting and signage and the redundancy needed to make egress safe and efficient in the event of a power outage (Building Officials and Code Administrators International, Inc. [BOCA], codes).

In healthcare facilities, secondary or battery-backup systems provide needed redundancy, while exit lighting and signage facilitate the timely flow of exiting traffic. Signage should be above and outside doorways, and lighting should illuminate traffic paths. Life-supporting equipment power must be supplied through a special-purpose, hospital-grade receptacle (NEC 517.13). Computer equipment must reside on an isolated ground so that its reference is isolated in the building (NEC 517.11).

▶Chapter Review

1. Which NEC section describes how to size wires for conduit, as well as the conduit color-coding schema?
   a. 230
   b. 250
   c. 314
   d. 517

2. To which NEC article would you refer to differentiate sheet metal gutter and a nonmetallic auxiliary gutter?
   a. 366.22
   b. 366.2
   c. 250
   d. 90

3. NEC Sections __________ and __________ discuss box sizing in cubic inches, as well as the sizes needed for varied 12 AWG and communication wires.

4. Which NEC table helps facilitate metal box sizing and cubic-inch determination?
   a. Table 6, Chapter 9
   b. Table 314.16(A)
   c. Table 2, Chapter 9
   d. Annex B

5. NEC Sections 314.16(B)(1) to (B)(5) provide guidance when deciding when to count a grounding conductor and a neutral conductor plus device sizing.
   a. true
   b. false

6. NEC Sections __________ and __________ provide(s) guidance for conduit sizing, number, and location.

7. When installing wires into metal raceways and pull, junction, and panel boxes, refer to NEC Sections __________ and __________ for insulator information.

8. Electricians who are repairing drywall and plaster and determining the precautions governing the spacing of boxes to drywall should consult NEC 314.21.
   a. true
   b. false

9. An electrician who must make a box legal with ____________ because the box sits too far in a wall or is overfilled with wires should refer to NEC ________.
   a. extension rings, 314.22(A)
   b. external grounding, 314.22(B)
   c. extension rings, 314.22(B)
   d. a and c

10. NEC Section(s) ________ outline(s) varied ways to support boxes with straps, boxes, and fasteners.

11. To differentiate single- and three-phase services, including where and how they must be installed, refer to NEC Section(s) ________.

12. NEC Sections 230.24(A) through (D) identify height requirements for service drops.
   a. true
   b. false

13. Point-of-attachment requirements appear in which NEC sections?

____

____

____

____

14. NEC Section(s) ________ discusses which authorities have jurisdiction in certain areas.

a. 80

b. 90

c. 100

d. none of the above

15. To learn how to support and protect service cables, an electrician should reference NEC Section 367.14.

a. true

b. false

16. Where in NEC would an electrician refer for information on the overhead services of commercial buildings and the different ways of mounting and wiring those services?

____

____

17. The reasons for enclosing or guarding service equipment appear in NEC Section(s) ________.

18. Which NEC section(s) describe(s) the sizes of service disconnects and how those disconnects are treated for installation?

____

____

19. What switch type can be used to control a 120V motor drawing 10 amperes?

a. 15-amp AC

b. 20-amp AC

c. 20-amp DC

d. all of the above

20. The best way to ground computer and data-processing equipment is to bond the receptacle ground to the metal raceways directly.

a. true

b. false

21. The metal yoke of a receptacle is common to the screw terminal for ground unless the receptacle is an isolated ground receptacle.

a. true

b. false

22. What type of conduit must be used in cinder fill?

____

____

____

____

23. NEC Table 5 contains which of the following types of information?

a. conductor temperature
b. conductor volume
c. box construction
d. conduit strength

24. Give two reasons to use RNC under a sidewalk.

____

____

____

____

25. Determine the neutral current in a 208Y/120V four-wire system when the phase currents are 20, 40, and 60 amperes through Phases A, B, and C, respectively.

____

____

____

____

26. An underground location is considered a wet environment.

a. true
b. false

27. When may a surface raceway extend through a partition?

____

____

____

____

28. When are the top and bottom elements of a typical hot-water heater used?

a. both are always used
b. top is used when the bottom fails
c. bottom is used when the top fails
d. only one is used at a time

29. How do a panelboard and a switchboard differ?

____

____

____

____

30. In reference to switchboards and panelboards, what does the phrase *condition one* mean?

____

____

____

____

CHAPTER

6 Commercial Wiring Techniques

CHAPTER SUMMARY

Commercial wiring is a multifaceted proposition that requires the electrician to plan, calculate, and execute efficiently, all the while keeping safety concerns paramount. Once installation plans are in place and electrical loads are calculated, the electrician can commence with mounting panels and boxes, manipulating conduit, and placing outlet and receptacle boxes. From feeder to panel and from panel to receptacles, switches, devices, and equipment, commercial branch circuits are constructed much like they are in residential environments. With a keen understanding of all tools at hand and a respect for each installation's unique underlying hazards, commercial electricians are well equipped to service customer needs.

▶ Learning Objectives

- Calculate branch and total loads
- Contrast conduit mounting methods
- Explain how to bend conduit to create a circuit
- Identify the uses of different types of fittings
- Discuss lighting circuits
- Outline means of working with outlets and receptacle boxes
- Map the process for installing branch circuits
- Compare types of conduit
- Describe NEC guidelines for raceway selection
- Identify procedures for special switches
- Discuss the approach to underground installation
- Explain how to work properly with lighting and signage

▶ Key Terms

conduit body
connected load
contactor
duty cycle
finish work
flexible metallic conduit (FMC)
hickey
knockout
liquid-tight flexible metallic conduit (LFMC)
offset
paddle fixture
reamer
rigid nonmetallic conduit (RNC)
saddle bend
strap
switchgear
time switch
trough
waterproof
watertight
weatherproof
wireway
yoke

When planning a generic commercial electrical system, the first step is calculating the electrical loads of things like lights, receptacles, and equipment. According to NEC Article 220, **connected load** is the load at installation. When paired with connected load, **power density**, measured in VA per square foot or watts per square meter, provides an easy way to determine power

Hints and Tips
When planning a commercial electrical system, create a table of job requirements that includes things like occupancy type and fixture types, numbers, and power requirements. Such a table facilitates both accurate job bids and efficient installation.

Hints and Tips
For track lighting, calculate power density using 75W per foot of track length (NEC 220.43[B]).

Did You Know?

Air conditioning is the environmental-control system that typically requires the most power. As a result, it appears in all heating, ventilating, and air conditioning (HVAC) calculations. Typically, it requires more energy than many heating systems, some of which only require an igniter, control logic, and a pump or fan.

requirements for lighting. NEC Table 220.12 is helpful as it provides unit loading information.

▶ Calculating Electrical Loads

Branch circuitry develops from a tally of all loads in all areas in each type of occupancy. Lighting is a load in nearly all commercial installations; other loads must be determined according to project. Most commercial lighting is a continuous load; it is energized 100% of the time. Few other loads or appliances have the high-**duty cycles** of lighting. Most buildings have environmental control in the form of air conditioning and/or heating, while some, especially those that manufacture, distribute, or use computer-based equipment, have humidity control.

Most air-conditioning systems have three motors. To calculate load requirements for the two single phase motors that run a system's evaporator and condenser, multiply the rated current by the rated voltage. To determine load for the third motor, a three-phase motor that operates the compressor, use the formula $VA = \sqrt{3} \times 208 \times \text{current}$.

Most commercial installations have one or more receptacle circuits. For loading, each **strap**, or **yoke**, must be calculated at 180W minimum (NEC Section 220). When a receptacle is placed to accommodate a specific load, that load may replace 180W in the calculation. The minimum requirement for signage circuits is 1200W (NEC 220.14[F]).

Hints and Tips
In load calculations, allow a buffer by multiplying lighting figures by 125%. This allows for some headroom, ensuring that conductor ampacity limits are not exceeded.

Did You Know?

In most commercial environments, lighting consumes the largest percentage of energy.

Hints and Tips
For detailed information on air-conditioning load requirements, consult NEC Article 440. For guidelines on motors and appliances other than air conditioners, see NEC Articles 430 and 422.

Hints and Tips
For each branch circuit, plan on 13 receptacles maximum (13 receptacles at 180 VA = 2,340W / 120 V = 19.5 A).

Did You Know?

Electrical metallic tubing (EMT) raceways are usually used in installations with metal studwork, while rigid metallic conduit (RMC) is usually used in masonry installations. This is because EMT is not sufficiently robust to withstand the pressures associated with masonry installations, although in more temperate climates EMT may be used because freezing temperatures are not an issue.

▶ Mounting Panels and Boxes

In general, commercial panels and boxes are mounted in one of two ways: (1) to metal studwork, which is analogous to residential 2 × 4s, or (2) in or on concrete walls.

Working with Metal Studs

Most manufacturers of metal studs punch holes in their studs for running conduit. When such holes are insufficiently sized or inappropriately placed, create some using a laser leveling device or a long bubble level and a hole saw or punch. Mount boxes such that their surfaces are level and flush with finished walls. In other words, when the cover plate is installed, it should contact the edges of the box and leave no gap with the wall. When working in unfinished environments, consult with general contractor's or customer's plans and specs to determine where finished walls will reside. When screwing a box to metal studs, use small clips to hold the box.

Job Connection

Mario was proud of his eight-member crew. Together, they had completed a commercial installation ahead of schedule and according to budget. During the final morning walk-through, though, the customer expressed displeasure with the number of receptacles he had requested. Wishing to please his customer and accommodate future expansion, Mario asked his customer to place small Xs on walls wherever he wanted receptacles added. The electrical team, Mario said, would assume the added cost.

After lunch, Mario reassembled his team to install 12 more receptacles over three branch circuits. When the customer returned at the close of business, he was so pleased with the work, completed quickly and above code, he awarded each team member a financial bonus. In addition, he asked Mario to bid on an upcoming project. Those few extra receptacles cost Mario and his team some added time and effort, but they paid many dividends in customer service and ongoing business.

ALERT!

When punching holes in metal studs, smooth any rough studwork edges with a **reamer** or file. While such edges do not jeopardize conduit installations, they can give painful cuts. Also, always keep the pointed ends of screws away from fingers as they will drill into skin as easily as drywall, if not more easily.

Did You Know?

Per Building Officials and Code Administrators International, Inc. (BOCA) requirements, the diameter of a conduit hole can only be within one-third of the width of the stud in which it resides.

When a commercial environment has wood framing, the techniques for metal studs apply. In residential installations, proper procedure includes drilling holes through members and creating notches in member faces to run Romex along walls. The same practices apply to commercial installations, although RMC requires no nail guards in front of notches (NEC 300.4). To satisfy proximity requirements when mounting boxes in commercial environments, as well as to make attachments more rigid near boxes, it may be necessary to insert a crosspiece between two studs (NEC 300.4[B][2]).

ALERT!

When using NM in metal studwork, protective grommets and bushings must be used as listed.

Hints and Tips
Use scraps of 2 × 4s to make stud crosspieces.

Mounting in Cement Environments

Electricians attach panels and boxes to concrete blocks just as they attach them to wood or metal; they simply use different parts. In poured concrete applications, as opposed to modular or premade concrete blocks, boxwork installs on concrete framing and conduit runs outside concrete boundaries. Special concrete boxes have little metal tabs that set into the concrete to improve the mount.

Hints and Tips
When mounting in cement environments, be sure to consult with your local inspector for applicable guidelines. The most successful masonry installations feature close collaboration between mason and electrician. To facilitate electrical work, ask masons to indicate when they are approaching courses where wiring and related elements will be installed.

Learning to Manipulate Conduit

A key task for the successful electrician is to learn to manipulate a cornerstone of electrical work: conduit.

Bending Conduit

Bending conduit in any direction is simple. Bending it to meet functional and aesthetic standards is an art form. While conduit may be purchased in prebent lengths, the pneumatic benders that create such lengths are typically found only in production shops. Instead, most conduit is bent on jobsites, which decreases part cost but increases labor time and cost. Generally, electricians use a tool called a conduit bender or **hickey** to bend conduit (Figure 6.1).

Figure 6.1 Conduit Bender or Hickey

Hints and Tips

To bend conduit at desired points, properly align that conduit in the hickey. First, place the hickey on a clean, flat surface. Then, apply foot pressure while pulling the hickey stick in a firm, continuous motion. Stop pulling when the back end of the conduit sits fully in the hickey's groove. Practice conduit bending before doing so on a job, and follow manufacturer instructions to accommodate hickey differences, particularly when executing complex bends. When bent within specifications, conduit can bend back into place.

When using a conduit bender, the angle of hickey-stick movement mirrors that of the conduit when bent. Manufacturer documents often outline the appropriate measurements for conduit benders (Figure 6.2); defer to make and model specifics. Measurement A is the length of the stub, the straight part after the bend, while measurement B is the length of the bend itself. In addition to requiring no boxes, bends use less conduit than L routes (Figure 6.3). For example, one piece of 0.75-inch conduit bent to a 5.0-inch radius requires 3.25 inches less of conduit than two pieces of conduit used to make an L connection.

Hints and Tips

When mounting a box a few inches above a horizontal run, leave a stub of conduit to facilitate box connection. Back the hickey off the combined length of the stub and the bend radius.

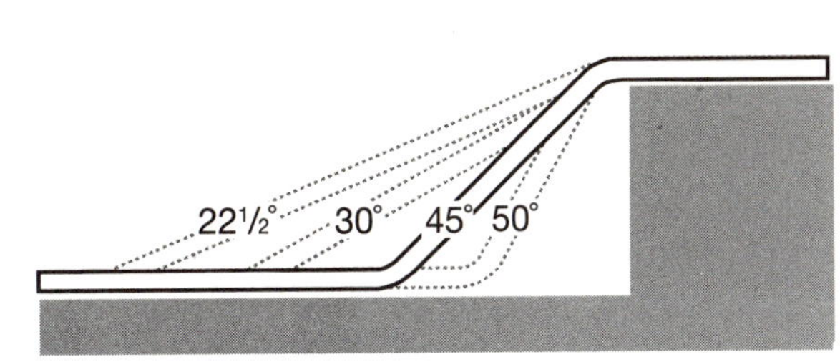

OFFSET DEPTH IN INCHES	DEGREE OF BEND							
	22½°		30°		45°		60°	
2"	5¼"	⅜"						
3"	7¾"	9/16"	6"	¾"				
4"	10½"	¾"	8"	1"				
5"	13"	15/16"	10"	1¼"	7"	1⅞"		
6"	15½"	1⅛"	12"	1½"	8½"	2¼"	7¼"	3"
7"	18¼"	1 5/16"	14"	1¾"	9¾"	2⅝"	8⅜"	3½"
8"	20¾"	1½"	16"	2"	11¼"	3"	9⅝"	4"
9"	23½"	1¾"	18"	2¼"	12½"	3⅜"	10⅞"	4½"
10"	26"	1⅞"	20"	2½"	14"	3¾"	12"	5"

Figure 6.2 Conduit Bending Measurements (*Courtesy IDEAL*)

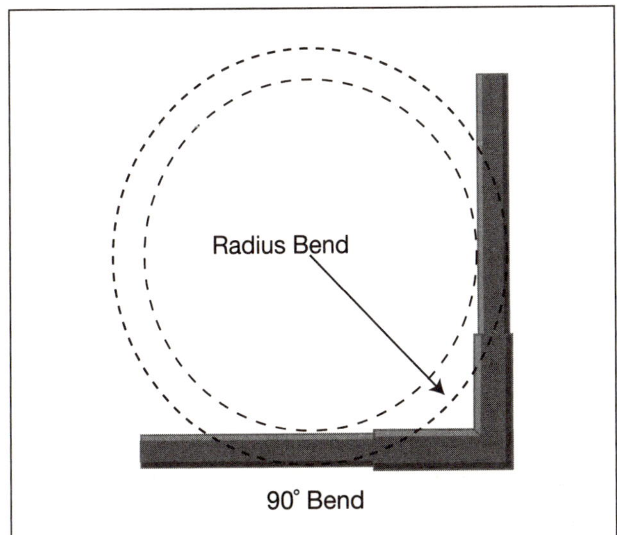

Figure 6.3 Conduit Savings by Radius Bend

On the common end, conduit bends run the gamut from 45- and 90-degree angles to wall-hugging supply lines that bend a few inches from walls to box openings. Other, less common types include the **saddle bend**, in which a piece of conduit bends out and then immediately in to cross a perpendicular piece of

Hints and Tips

Use the knowledge that bends save conduit to precut conduit for installations. Precut conduit is particularly convenient when the weather is poor.

> **Hints and Tips**
> Before cutting conduit, ensure blades are in good condition and replace them as needed. Secure the conduit in a vise, and fully mark cuts around the conduit's circumference. Measure twice, and cut once, slowly and deliberately. To cut cleanly, support the waste end of conduit while cutting. When using hacksaws, choose blades with 24 teeth per inch.

conduit; parallel offset, in which several runs of conduit follow the same parallel path; and quad, in which a piece of conduit crosses two parallel obstructions, like hot and cold water lines (Figure 6.4).

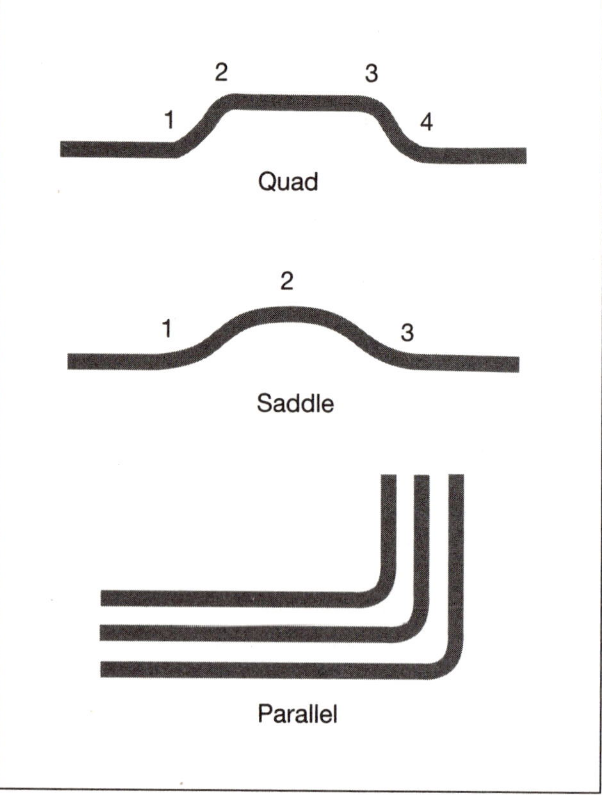

Figure 6.4 Saddle, Parallel, and Quad Bends

Cutting Conduit

Generally, electricians cut conduit with hacksaws or plumber's pipe cutters. Pipe cutters are the easier option, especially for RMC and other thick-walled raceways. For polyvinyl chloride (PVC), hacksaws are the preferred choice, because the flexible PVC can be more easily managed in the field when not using a power tool. Some electricians cut large-diameter PVC with reciprocating saws or a portable band saw and use rotary tools on small-diameter materials. A Dremel tool may work better and faster than a hacksaw, but it requires power and offers no guarantee of a planar cut.

> **ALERT!**
> The conduit-cutting process can create edges or shards that can slice flesh as well as conductor insulation. Therefore, be sure to smooth any rough edges in order to be compliant.

> **Hints and Tips**
> Before proceeding to a cutting station, mark all conduit ends to indicate required thread depth.

Threading Conduit

The ends of RMC are often threaded by machine or hand to facilitate connections. Hand threading involves inserting conduit into the tapered side of a lubricated die and turning until the conduit is threaded to the desired depth.

Hints and Tips
When hand threading conduit, oil the die while turning. Every half-turn, move the die backward slightly to allow any steel chips to fall or wash away. Do not exceed the desired thread depth, because extra threads are contaminant pathways. Once threading is complete, check inside the conduit for sharp edges. If edges exist, ream the conduit again to smooth them.

▶ Using Fittings

Conduit is not just inserted into holes. Instead, it affixes to boxes or other apparatuses via fittings. In electrical work, fittings provide solid mechanical, or physical, connections between two items. Locknuts and bushings, for example, secure conduit to boxes. When threading is not an option, such as is the case with thin-wall or nonmetallic conduit, fittings are required on both sides of conduit breaks. Set screws or compression rings secure most fittings; combination couplings connect rigid and flexible conduit pipes.

Did You Know?

As fittings, locknuts provide security, while bushings provide smooth paths that safeguard conductors from damage arising from sharp box or conduit edges. When metallic and installed properly, both locknuts and bushings provide electrical connectivity.

Hints and Tips
Work quickly with PVC connections. Once pieces are mated, they cannot be taken apart.

ALERT!

PVC parts are welded together chemically. The solvents and adhesives that are required render the process hazardous. For details on ensuring safety, consult the solvent material safety data sheet (MSDS), but at minimum, provide adequate ventilation.

For metallic conduit, fittings must continue the ground path from grounding point throughout the conduit run. Pieces of PVC, once they are cut, reamed, and cleaned with a solvent, are joined with a noxious glue.

▶ Applying Outlet and Receptacle Boxes

In commercial installations, boxes house switches and receptacles as well as serve as pull and junction boxes. A one-gang box, for example, could hold two receptacles, two switches, or one receptacle and one switch. It could also simply serve as a junction box. In all cases, the conduit and box remain the same. Only the box's cover differs.

Electricians must be able to tailor boxes and box covers to applications. Will a box be used indoors or outdoors? In a hazardous location? Will it support a

Hints and Tips
When installing boxes outside, check with local authorities for their interpretations of guidelines.

certain weight? Mount in concrete or masonry? Answers to questions like these help drive box selection.

The environment is just one factor by which boxes are chosen. **Weatherproof** equipment is impervious to weather, including rain, snow, ice, wind, and ultraviolet radiation. It is not **waterproof** or even **watertight**. Watertight equipment allows no water in, while waterproof equipment is impervious to water, even when submerged. Therefore, watertight equipment is desired in refrigerated areas, for example, while waterproof equipment is desirable in areas prone to flooding. In most cases, weatherproofing is adequate protection for boxes installed outside.

Like the environment, a location's hazard level helps drive box selection. Class III locations, the least hazardous, may experience periods of airborne flammable particulate matter. While sawdust from a sawmill is particulate matter, it is not ignitable and generally carries no risk of explosion. Like Class III locations, Class II locations have airborne particulate material but at sufficient density to combust and risk explosion and/or fire. In Class I locations, flammable gas may be at sufficient density to explode in the event of an electrical arc. In a Class I location such as a paint booth, raceways and boxes must be sealed and rated as explosion-proof.

In general, outlet and receptacle boxes are made of one of three materials: (1) pressed steel, (2) cast iron or brass, or (3) nonmetal. Accessories help attach boxes to different structural elements. Tabs, or ears, for example, set in concrete. Some boxes mount flush with surfaces, while others mount on surfaces. Square boxes typically mount flush.

Did You Know?

Square boxes, available in 4-inch and $4\frac{1}{2}$-inch sizes and depths from $1\frac{1}{4}$-inch to $2\frac{1}{8}$-inch, are the most popular versions today due to their increased volume. Octagonal and round boxes, used when boxes are exposed for lighting and switching purposes, are sometimes preferred in residential installations given their visual appeal.

Did You Know?

Explosion-proof hardware is impervious to penetration by noxious or flammable vapors.

Hints and Tips
Allocate box space properly. Recall that box capacity is based on number of conductors, conductors' gauges, and the presence or absence of devices, or straps, in the box (NEC 314.16). For the proper use of pull boxes, used to draw conductors through complex conduit bends, see NEC Section 314.28.

All boxes, except nonmetallic ones, have accessory extension rings on their faces to increase volume, which enhances the boxes' capacity to accommodate conductors. Given their makeup, nonmetallic or nonconductive boxes are suited to moist locations, but they must be approved (NEC Section 314.15). Similar to boxes in moist environments, boxes that support fixtures have special ratings (NEC 314.27). A ceiling fan, for example, requires a box that is listed as able to support the fan's weight. **Paddle fixtures** are used for some heavy appliances and ceiling fans (Figure 6.5). In addition to a box, a paddle fixture includes a horizontal brace and the hardware needed to install that brace between two ceiling joists.

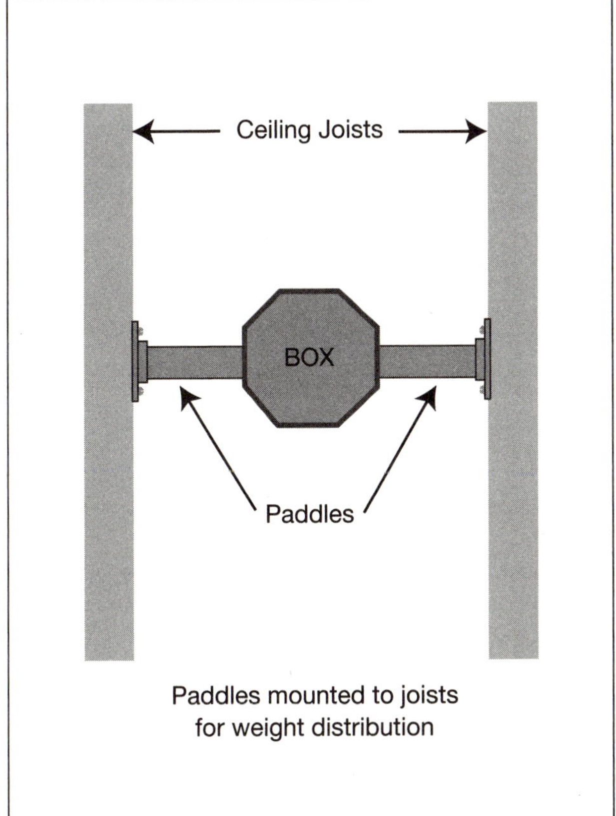

Figure 6.5 Paddle Box

▶ Installing Branch Circuits

In most ways, commercial branch circuits are constructed much like they are in residential environments. Work commences from feeder to panel(s) and from panel(s) to receptacles, switches, devices, and equipment. It is the high-gauge wire of commercial installations that poses unique installation challenges. For example, bending four-aught (0000-gauge) wire can be challenging, but wrapping it around a lug is even more difficult. Fortunately, many lugs for high-gauge wire adopt a straight-through design in which wire is simply inserted through a collar that is tightened by screws to achieve physical and electrical connections.

To connect branch devices, conduit must be cut, bent, reamed, and interconnected precisely. On the way to its destination, nearly all commercial conduit runs in some form of raceway. As discussed in an earlier chapter, raceway is a channel that carries conductors. In residential installations, Romex is typically used to connect point-to-point. In commercial terms, Romex has its own raceway.

In both commercial and residential installations, all metallic elements, including conduit, boxes, panelboards, **switchgear**, and other appliances, must be grounded. Devices can be grounded in two ways. In the first, an equipment grounding conductor runs from equipment to a grounding terminal in the panelboard where the circuit conductors originate. In the second

Did You Know?

Some of the thickest wires in commercial installations weigh tens of pounds per foot.

Hints and Tips
Local rules and regulations sometimes govern the use and installation of raceways. Therefore, check with local authorities before raceway use.

Did You Know?

EMT is the lightest-gauge raceway type.

grounding method, conduit that feeds an appliance or a device acts as an equipment grounding path (NEC 250.118). Conduit like PVC tubing, which conducts no electricity, cannot be used this way. Instead, the raceway must contain an equipment grounding conductor as well as current-carrying ones.

Raceway, like conduit, comes in different forms. Rigid metal conduit (RMC), for example, is thick-walled pipe that typically runs conductors through or under concrete but not in or under cinder concrete or fill, especially in damp or corrosive areas (NEC 344.10[C]). Given its robust construction, RMC may also serve as an equipment grounding conductor. Traditionally connected using threaded fittings, RMC bends to navigate around structures and appliances. It may not, however, exceed 360° total bend before using a pull box (NEC 344.24–26). Being fairly rigid, RMC must be supported within 3 feet of its source and destination and at every 10 feet along its length (NEC Article 344).

RMC looks like a second raceway type, electrical metallic tubing (EMT), but the two are not interchangeable. Unlike RMC, EMT has thin walls and lacks threading, relying instead on couplings and connectors for joining. In moisture-prone areas, compression fittings provide watertight seals. In other areas, set screws render the EMT rigid.

EMT may not be used where it is subject to physical damage, places like corrosive areas, under or through cinder concrete, or in cinder fill (NEC 358.111). Given its thin walls, which make it easier to bend than RMC, it may not be used to support fixtures or boxes. Instead, it is appropriate for concealed or exposed areas (NEC 358.10) or as an equipment grounding conductor when it has special fittings marked "ground" (NEC 250.118). Like RMC, EMT must be supported at least every 10 feet along its route, as well as within 3 feet of a box on either end (NEC 358.30).

Hints and Tips
Before making RMC connections, coat the threads with a noncorrosive, electrically conductive compound. To save time and ease navigation around structures, purchase prebent RMC. Support RMC conduit by running it through joists or studs. Ream the ends of RMC to remove any rough edges that could nick or cut conductors. Whatever raceway is used, render all runs level, straight, and plumb.

Hints and Tips
Support FMC at closer intervals than those suggested by the NEC. Wires pull through installations most easily when FMC is well supported.

Hints and Tips
Although FMC is easy to bend, do not exceed its published bend radius, which is the same as that for more rigid conduit (NEC 384.24–26).

Hints and Tips
LFMC is subject to the same requirements as FMC. For rules governing all types of FMC, see NEC Sections 348–50.

Flexible metallic conduit (FMC), a third raceway type, is used where rigid conduit would be inconvenient, difficult, or impossible or connectivity will be established with a vibrating entity. The commercial equivalent of Greenfield, a premade FMC without conductors, FMC is used most commonly to establish connectivity to motors and to lighting fixtures above drop ceilings. To avoid having a separate equipment grounding conductor through conduit, it can sometimes serve as an equipment grounding means when it is less than 6 feet long and carries fewer than 20 amperes (NEC 348.60, 250.118).

FMC must be at least $\frac{1}{2}$ inch, except when it is used with luminaries, in which case it may be $\frac{3}{8}$ inch (NEC 348.20[A]). At $\frac{3}{8}$ inch, it can run a maximum of only 6 feet long, however. In any run, FMC must be supported within 1 foot of a box, as well as every 54 inches between boxes or devices (NEC 360.11). FMC connects to boxes with a flexible connector and to rigid conduit via a combo coupling. When connected to an equipment grounding system, FMC requires a bonding jumper. When an installation must be watertight, a liquid-tight version of FMC, called **liquid-tight flexible metallie conduit (LFMC)**, is used.

Hints and Tips
When using FMC as an equipment grounding means, be sure to use fittings marked as such.

Hints and Tips
Verify the integrity of all FMC connections. It is far more challenging to make watertight connections truly watertight, for example, than it is to make those connections simply look good.

▶ Applying Rigid Nonmetallic Conduit

Rigid nonmetallic conduit (RNC) is made of the PVC plumbers use to route water because, like its plumbing counterpart, RNC uses a solvent to create permanent watertight seals. Overall, RNC is allowed in most areas: behind walls, in concealed areas, in damp locations, and underground (NEC 352.11). It is disallowed from hazardous locations, as fixture support, and where temperatures may exceed 50°C (NEC 352.12). Unless it conforms to NEC 518.4 and NEC 520.5, it may not be used in places of assembly, such as theaters, churches, and concert halls.

Like some other conduit types, liquid-tight versions of RNC are available. As a nonmetallic raceway, liquid-tight FNC (LFNC) or electrical nonmetallic tubing (ENT) is like plastic Greenfield. It is easy to work

ALERT!

RNC (rigid PVC) for electrical use must be listed as such. Replacing RNC with PVC plumber's pipe for electrical work violates NEC Article 352.

with, can be bent by hand, and can be used in branch circuits as well as low-voltage wiring.

To accommodate differences in installations, RNC comes in two configurations: Schedule 40 and Schedule 80. Schedule 40, also called Type EB or Type I, is a general-purpose conduit suitable for use indoors, above or below ground, and in areas without risk of physical damage. As a more robust option, Schedule 80 RNC, also called Type DB or Type II, may be used where physical damage is possible. At the expense of its inside volume, the outside wall of Schedule 80 RNC is thicker than that of Schedule 40 RNC. According to National Electrical Manufacturers Association (NEMA) standards, all RNC must be marked during manufacture to include manufacturer name, length, material, and conduit type.

In general, most single runs of RNC may be no more than 10 feet (NEC 362.11), although the conduit is available in longer lengths. Because RNC has a high coefficient of expansion, long runs may use expansion couplings with O-rings to allow run flexibility. RNC must be supported at closer intervals than some of its more rigid counterparts (NEC Table 352.30). For conduit $\frac{1}{2}$ inch to 1 inch in diameter, support intervals range from every 3 feet; for 6-inch diameter conduit, those intervals go up to 8-foot.

Did You Know?

ENT comes in yellow (communications), blue (power wiring), and red (emergency wiring) variations.

▶ Studying Conduit Bodies

To be able to pull cable easily when routing conduit, most electricians restrict bend total in a run to 270°. Sometimes, however, electricians must exceed the maximum number of bends allowed in a run, which is 360°. When they do, they need a way to be able to pull cable past difficult points. **Conduit bodies** provide access to conduit conductors without need for junction boxes. Such bodies bend on their own. In addition to straight-through conduit bodies, called Type C, conduit bodies are available in a 90° bend, called Type L; a three-conductor tee connection, called Type T; and a Type X version that allows four-way crossing of conductors (Figure 6.6).

Hints and Tips

When installing conduit bodies, include all bushings needed to keep any sharp edges of conduit connections from damaging the conduit's inner insulation and conductors (NEC 300.15). When installing in moist, damp, or corrosive areas, use bushings designed to maintain a waterproof environment.

Figure 6.6 L, T, and C Conduit Bodies

When used properly, conductors are typically pulled from one end of a conduit body, coiled on the floor, and then pulled through the conduit body's exit side. Once all conductors are pulled through, the gasket and cover plate are centered and secured so the next task, wiring the devices and outlets, can commence.

▶ Installing Branch Circuit Switches

In addition to using switches like those used for residential lighting, commercial establishments allow employees to control branch circuits from one place using a series of switches mounted in a panel box. In simple terms, a panel box with breakers is considered switchgear if it handles over 600V. Otherwise, it is called a panel box. Neither panel box, metering device, nor switchgear can be part of a raceway (NEC 404.3). Similarly, no enclosures can be used as junction boxes (NEC Articles 210–20). Switchgear that fails must be replaceable without shutting down other lines, except those inside the switchgear.

Commercial panel boxes can offer direct control of lighting and fixturing, as well as other electrical devices and loads. Boxes used for routine switching loads must be easily accessible, and switchgear should be mounted 6 feet 7 inches high from finished floor

> **Hints and Tips**
> When installing switches in commercial installations, use wire size to determine the cubic inches needed in a box. In addition, ensure that the yoke of each device is grounded from panel to box and from box to switch.

ALERT!

Failing to observe and follow switchbox stickers that state "Do not turn off" or that hold breakers in place compromises efficiency as well as safety.

level to allow all markings to be visible from a standing position (NEC 404.8 and 240.24). For some devices or pieces of equipment, switching may be installed on or within eyesight of units (NEC Article 100). Switches at greater heights that require hook sticks for operation may be installed at ceiling height as long as they are easily accessible and carry clearly marked "On" and "Off" switches.

After electrical service has been roughed in at a location, the tasks needed to complete job, called **finish work**, occur (NEC 404.9). Finishing, or constructing, boxes into masonry, drywall, plywood, or ceiling tile requires that boxes be flush with finished wall plates on installation surfaces (NEC 314). Wall plates made of metal, plastic, or any nonferrous material must cover the switch area as well as the opening (NEC 314.25). A switch's metal framing must be connected to an equipment grounding conductor by metal screws worked around clips (NEC 290.5). Plates must be at least 0.03 inch thick but not more than 0.04 inch thick (NEC 290.5). In general, all metal entities, including boxes, must be bonded to circuits' equipment grounding conductors. Connecting plastic devices to grounding systems requires equipment grounds with current-carrying conductors.

ALERT!

To ensure personal safety, always use a tester to verify that a unit has been disconnected from a power source. Additionally, per Occupational Safety and Health Administration (OSHA) requirements, always don safety glasses when working around hazardous voltage levels.

▶ Considering Special Switches

Commercial installations sometimes require that devices mount on outside building surfaces, such as on rooftops near heating systems or on exterior walls near air conditioners. To ensure proper functioning, such switches and/or receptacles must reside in surface-mounted, concealed, weatherproof boxes with weatherproof plates (NEC 404.4, 406.8) With an air-conditioning unit, for example, a receptacle must be close to the unit for service purposes, as well as protected by a ground fault circuit interrupter (GFCI) via a GFCI receptacle or a GFCI-rated breaker (NEC 404.4).

Many commercial applications use **time switches** to control such equipment as parking-lot lights and advertising signage. In large installations where lights may exceed 250W, equipment timers must include a set of **contactors** (NEC 409). With 20 lights on a circuit, each at 250W, circuit wattage would be 5,000. Powered by a typical 240V system, the amperage would be 20.83. Rounded up, 21 amps is too much draw for a 20-amp breaker. A contactor used as a relay will draw much less than 20 amperes but control much larger currents. The setup is analogous to that of vehicle headlights. To keep the fairly high currents of headlights from a vehicle's cockpit, a low-power circuit is closed by a headlight switch that controls the higher power circuit between the battery and the headlights through a relay.

When such a setup occurs in a commercial environment, a manual disconnect must be accessible to building patrons and/or employees, except when general maintenance personnel are trained in the disconnect's operation. A switch's on/off position must be clearly engraved or otherwise permanently marked on the device so that all people can identify it regardless of natural spoken or written language (NEC 404.7).

▶ Instituting Grounding in Raceways

An electrician running wire in raceways or armored cable must ensure that the grounding conductor is never disconnected (NEC Article 250). Switch loops do not require a grounded conductor, because only one ground is required as long as it is sized to the largest conductor in that raceway. When inserting instances in existing locations, grounding conductors are not permitted in a separate raceway (NEC Article 300.20). For example, assume you are installing circuits in an existing building when you discover an old panel box lacking both grounds to the other, existing circuits and a separate, derived ground into the existing panel box. From the back panel to a ground inside or outside the building, you should run a separate ground sized to the largest current-carrying conductor. Inside the panel, you should separate the ground and neutral by adding a ground bar and removing the bonding screw.

As discussed previously, a grounding electrode system, which is often a ground rod, is used to bond metal objects to the earth (NEC Articles 250.50 and 250.52). All metal fixtures, like metal parking lot lampposts, must be grounded. In this case, the lampposts might be grounded by routing a grounding electrode conductor from the equipment grounding conductor connection in the lamppost junction box to the grounding rod. This is important because such fixtures, depending on their heights and the presence or absence of other tall structures nearby, such as trees, are highly susceptible to lightning strikes. A strong path to ground helps keep neighboring buildings and their occupants safe. Ultimately, the grounding electrode conductor must connect the posts of every lighting fixture on the premises to the grounding rod. The grounding electrode is attached first, and then it is married to the grounding rod through a weld or a rated clamp, often called an acorn clamp. A lug should be placed on the post end so the post can be replaced without destroying the ground connection.

▶ Manipulating Grounding and Other Boxes

The boxes used to mount devices and fixtures, like pull boxes, come in metallic and nonmetallic versions. Metallic boxes, which are made of galvanized steel, must be punched with holes to accept conduit termination bushings. **Knockout** punch kits remove the circular piece that makes a hole. To remove a prepunched

> **Hints and Tips**
> To avoid the effort of creating knockouts, purchase prepunched boxes.

Hints and Tips

If needed, use a small adapter to manipulate conduit from the surface where it is mounted to inside the box. This short, curved piece of conduit is called an **offset**.

pry out, hold a flat screwdriver against one of the small metal tabs keeping the pry out in the box. Holding the box securely, tap the end of the screwdriver with a hammer to break the tab. Twist the knockout off, leaving a hole. Make sure that the connecting piece went with the tab and is not on the box. Otherwise, it may be difficult to insert a bushing.

When working in corrosive areas, use nonmetallic boxes (NEC Section 314), which generally follow galvanized-steel-box procedures. Like their metal counterparts, plastic boxes typically feature knockouts. Unlike their metal counterparts, plastic boxes are usually stamped inside with their volumes.

To complete the fairly straightforward process of connecting conduit and box, follow the "external locknut, box, grounding washer, and bushing" (EBGB) procedure. First, insert the conduit through the external locknut (E). The locknut will either screw onto the conduit, or there will be a compression fitting that must be tightened to secure the locknut to the conduit. Next, insert the locknut through the hole in the box (B), place the grounding washer (G) over the exposed conduit end, and attach the bushing (B) by screwing it onto the locknut.

Did You Know?

Troughs are more popular today for low-voltage and telecommunications wiring than they are for electrical service (NEC 392.3[A]). Troughs more easily accommodate the frequent upgrades of telecom and low-voltage wiring because they are open-sided raceways.

Supporting Raceways

As stated earlier, raceways must be supported close to boxes as well as at regular intervals between boxes and/or devices. Straps are ideal solutions for single runs of conduit. Typically made of metal and available in several different styles, the simplest and most economical strap is a single-hole model that has a half-circle with a tab. While the electrician can exercise some discretion in tab choice, application type is sometimes a factor. A single-hole strap is ill suited to running conduit along a ceiling, for example. In contrast, a two-hold strap (a strap that is fixed on both sides of conduit), is appropriate for ceiling conduit because it supports the conduit from both sides and prevents fatiguing clamps from giving way easily.

For multiple runs of parallel conduit, a bus approach that uses two ceiling-hanging beam clamps is appropriate. In this approach, a stationary lower beam clamp has grooves or cutouts for conduit runs. Once conduit runs are placed in these grooves, an upper clamp is placed over the conduit and attached to the lower clamp.

Wireways, which have hinged covers, and troughs, which do not, are effective ways to organize large quantities of conductors (Figure 6.7). These options are often exercised when appearance is a factor or conduit

Figure 6.7 Organized Commercial Installation

conductors cannot be hidden or installed. **Trough** systems, which are typically supported along walls using shelf brackets, suspended hangers, or gussets, include provisions for 90°-, T-, and cross-connections, much like conduit.

> **Hints and Tips**
> When installing wire underground, follow the maxim, "Measure twice, dig once." Consult NEC Article 300 for general guidelines, as well as the requirements of local jurisdictions, which are governed by weather as well as geological considerations. At minimum, confirm that it is acceptable to dig in the targeted area. In addition to examining things like water, gas, and cable lines, identify the locations of items like septic tanks and buried concrete slabs.

▶ Completing Underground Installations

Running wiring underground is an expensive, multistep process that requires proper planning. The process

Hints and Tips

To get a pull-rope or string through a raceway to pull wires, attach fishing line or dental floss to a ping-pong ball or plastic fishing bobber. Using a shop vacuum at one end of a tunnel, suck the ball or bobber through the tunnel. Once the line is through the tunnel, use it to pull a more robust line through the tunnel. This more robust line will, in turn, pull the conductors through the tunnel. Should further expansion be needed, keep a pull-rope in the conduit.

typically involves creating a tunnel at a specified depth, populating the tunnel with conductors and a fill specified by local ordinances, and completing terminations at each end. In addition, tunnel access should be spaced along the run and, when possible, the chase should be several sizes larger than required to accommodate expansion.

▶ Accommodating Lighting and Illumination

When installing luminaries at most commercial and industrial sites, electricians must remember to check wire sizing for conduit fills and box fills as discussed in previous chapters. Doing so helps ensure that heat dissipates from tight areas.

Most commercial lighting electrical power runs through $\frac{1}{2}$-inch or $\frac{3}{4}$-inch EMT (NEC Table 1 Chapter 9), although many municipalities now require at least $\frac{3}{4}$-inch conduit for all commercial applications to allow for expansion. Commercial electricians have two basic wiring choices: (1) RHH/W and (2) THHN/THWN. Assume, for example, an electrician is using 14-gauge wire. With RHH/RHW, the electrician can install only eight wires/five wires in a piece of $\frac{3}{4}$-inch conduit. THHN/THWN can populate the same $\frac{3}{4}$-inch piece of conduit with 17 total conductors (NEC Chapter 9, Table 1). Conductor fill requirements differ because wires' insulation differs. RHH and RHW are less heat tolerant than THHN/THWN, so a given piece of conduit appears to carry two to three times as many THHN/THWN conductors as RHH/RHW conductors (NEC Chapter 9, Table 1).

Wires strands are marked for additional information (NEC Table 402.3). For example, 14-gauge wire is typically associated with a 15-amp breaker, while 12-gauge wire is associated primarily with a 20-amp breaker. In many cases, however, a multitap transformer will feed luminaries through 16-gauge wire, because wire often terminates inside a light fixture. As a result, a pigtail is limited to about 6 inches of 16-gauge wire. It is common for a 16-gauge conductor to have insulation with a higher temperature rating than its connecting wire.

Temperature tolerance increases closer to a light fixture because the fixture generates heat. Depending on lighting type, ambient temperature near the ball, including where the wiring is, may exceed the 90°C rating for the branch circuit conductors (THHN). NEC

ALERT!

Conduit that cannot dissipate the heat generated by electrons flowing through its conductors can be a fire hazard.

Article 402.6 shows the size of wiring used in such light fixtures as 16 AWG, as found in NEC Table 402.5, which discusses allowable ampacity for fixture wiring. Consider the amp draw of a T8 ballast running 35W bulbs. Ballast inspection should reveal that the ampere draw is about 1.75, well under the 8-amp rating for 16-gauge wiring. Because the lighting circuitry in a commercial environment is not user accessible, an 8-amp circuit is allowable. After all, it is not a branch circuit per se.

Grounding of light fixtures must be marked for easy viewing by the installer (NEC 250.119). The grounding terminal, used for grounding purposes only, is normally marked green, but sometimes it is marked green with a yellow stripe. Different colors mark conductors for different purposes (NEC 400.2). Black is a hot wire, white is a neutral wire, and green is a ground wire. Such markings are usually included for 120V or 277V fixtures and luminaries.

In NEC 402, Table 402.5 indicates that 12 AWG wire may carry 23 amps, which seems contradictory to residential guidelines that 12-gauge wire could only carry 20 amperes. The discrepancy arises because, on short runs typical in commercial environments, 12-gauge stranded wire is used for lighting. Such 12-gauge stranded wire can carry slightly more current than solid core wire due to "skin effect." Electrons tend to move a long a conductor on wire's outside peripheries, or skin. One conductor of solid stranded wire only has one skin, whereas a conductor of stranded wires has many (Figure 6.8). Many skins conduct more current.

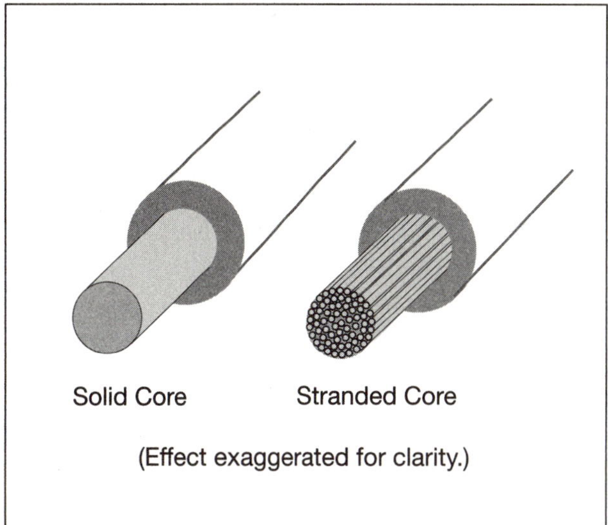

Figure 6.8 Stranded and Solid Wire

Job Connection

On Thursday, Wolfgang began installing some parking lot lampposts. By Friday, he had welded all grounds to the posts and expected to return after the weekend to set the lights and connect the wires. Unfortunately, he was greeted Monday by a tractor trailer caught on one of the completed posts. Because the post was wedged in the trailer, Wolfgang had to replace it, which entailed cutting the ground wire to the post, disconnecting the wires, and replacing the pillar post lights as well as the entire ground conductor. Had he taken the time to terminate the ground conductor properly to the post, he could have completed the repair much sooner and at much lower cost.

▶ Posting Exit Signage

In addition to the NEC, practicing electricians must know the National Fire Protection Agency (NFPA) regulations that apply to exit lighting (NEC Section 7.10). NFPA guidelines assume that, in the event of a fire or another emergency, the general public is at risk of confusion and will therefore be unable to exit the building efficiently. NFPA code describes where exit signs must and cannot be, for example, while NEC describes how to wire those signs. NFPA requires that exit lights be red or, in some cases, green, so exits are eye-catching and traffic moves outside as quickly as possible. An ideal exit sign is visible from any point in a building.

Emergency lights should be installed outside buildings to illuminate escape paths. All emergency signage should be directly lighted or backlit for visibility in the dark. Emergency lights connect to existing circuits, and emergency lights and signage alike must be connected to battery backup systems to ensure uninterrupted service. Emergency exit lights typically have multitap ballasts in addition to batteries. Some electricians connect to lighting or receptacle circuits provided that the receptacle circuits supply no life-supporting or other special-purpose equipment. Specific guidelines for emergency and exit lighting in healthcare occupancies appear in NEC 517.32(B) and 517.42(B); NEC Article 700 details the requirements of other establishments.

▶Chapter Review

1. For each of the following statements, provide an NEC reference.

a. Items under the exclusive control of the utilities are exempt from NEC
b. Current considerations for nonlinear loads
c. Requirements for motor disconnect switch
d. Installation requirements for EMT
e. Requirements for EMT
f. Requirements for services
g. Motor back EMF and relationship to short-circuit current
h. Dedicated space above electrical service equipment
i. Disconnecting means for permanently connected appliances
j. Length restrictions for liquid-tight flexible nonmetallic conduit
k. Appropriate use for Type AC armored cable
l. Equipment grounding conductor color
m. Illumination of electrical service equipment
n. Panelboard rating based on feeder capacity
o. Thermal protection of ballasts
p. Grounding of a neutral conductor
q. Attachment of receptacle cover to box
r. Grounding requirements for appliances
s. General load calculations
t. Appropriate use for type MC cable
u. Isolated ground connections at the panelboard
v. Overcurrent protection requirements for appliances
w. Overcurrent protection of conductors
x. Consideration of nonmetallic structural items as grounded

2. A right-angle bend is ____________ degrees.

a. 50
b. 90
c. 180
d. 45

3. The NEC is strictly for use in the continental United States.

a. true
b. false

4. A ____________ is used to smooth the inside edge of a piece of conduit.

a. hickey
b. sanding dowel
c. utility knife
d. reamer

5. RMC must be supported every ________ feet.

a. 5
b. 7
c. 10
d. 12

6. When connecting a motor subject to vibration, what type of raceway should be used?

a. RMC
b. DMC
c. FMC
d. EMT

7. Metallic boxes that are typically surface mounted are

a. octagonal and square.
b. square only.
c. round only.
d. octagonal and round.

8. Which of the following statements is true?
  a. Waterproof conduit will leak before watertight conduit.
  b. Steel and aluminum conduit may be used in corrosive areas.
  c. RMC and EMT may be used where damage could occur.
  d. Weatherproof conduit protects against rain, snow, and sun.

9. Nonmetallic boxes are available with extension rings to increase usable volume.
  a. true
  b. false

10. All electrical energy in a commercial installation must be controlled so it can be switched off with
  a. one master switch.
  b. a maximum of six switches.
  c. permission of the owner only.
  d. a key.

11. A conduit bender cannot be used to make offsets; offsets must be purchased from a distributor of electrical equipment.
  a. true
  b. false

12. The amount of light required for different occupancy types is
  a. at the building owner's discretion.
  b. dependent on local codes.
  c. specified by NEC.
  d. b and c

13. Heat dissipation is not a concern with fluorescent lighting.
  a. true
  b. false

14. In a Class ______________ hazardous area, gas vapors are the primary concern.
  a. I
  b. II
  c. III
  d. all of the above

15. When a grounding conductor is attached to a box, a(n) ______________ must be used.
  a. green screw
  b. grounding grid
  c. aluminum electrode
  d. grounding fitting

16. Electricians should use cable ties to support new LV wiring using existing conduit.
  a. true
  b. false

17. A ______________ mount is a common way to hang conduit from a ceiling or beam.
  a. traverse
  b. transverse
  c. trapeze
  d. trapezoidal

18. When mounting conduit to a ceiling, the least preferable clamp type is the
  a. two hole.
  b. solid body.
  c. single hole.
  d. none of the above

19. A(n) _____________ is a type of pull box that allows cable to route through more than 360° but does not allow for future expansion.

a. octagonal box
b. plastic box
c. conduit body
d. meter base

20. To improve ventilation in hot or damp areas an electrician should leave unused panelboard spaces open.

a. true
b. false

21. Instead of using a T-connection, a bend

a. uses less conduit.
b. is easier to construct.
c. requires fewer parts.
d. all of the above

22. The maximum number of receptacles on a commercial branch circuit is

a. 13.
b. 21.
c. 36.
d. 48.

23. Exit signage requires a receptacle capable of handling _____________ watts.

a. 100
b. 500
c. 750
d. 1,200

24. Relative to Schedule 40 PVC, Schedule 80 PVC has

a. thinner walls, a smaller internal volume, and is designed to be used in areas free of physical damage.
b. thicker walls, a larger internal volume, and cannot be used where physical damage may occur.
c. the same size walls but a smaller inside diameter.
d. thicker walls, a smaller internal volume, and may be used where physical damage may occur.

25. A typical commercial high-volume air-conditioning system has

a. three motors: one three phase and two single phase.
b. two motors: one three phase and one single phase.
c. three single-phase motors.
d. one motor that drives all components.

26. When must an electrician request an inspection?

____

____

____

____

27. Identify three safety tips for wiring a hot panel.

____

____

____

____

28. Branch circuits must be loaded equally in terms of wattage per unit area.

a. true

b. false

29. In a Class ________, Division ________ location, in unusual conditions dust may present an explosion hazard.

a. I, I

b. II, I

c. II, II

d. III, IV

30. In terms of fixture wiring, the NEC does not maintain requirements for ampacity. It is assumed that the manufacturer has had the fixtures testing as part of the UL-approval process and therefore no other requirements apply.

a. true

b. false

CHAPTER

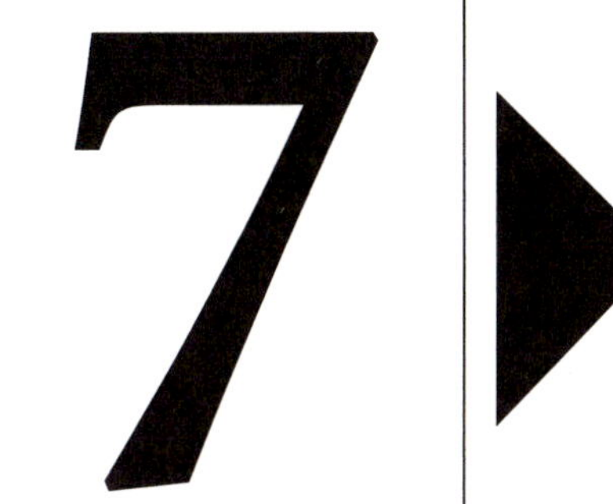

Protection at the Office

CHAPTER SUMMARY

Whether installations are residential or commercial, safety is a cornerstone of successful electrical work. A circuit in any realm that is vulnerable to shorts risks hazard to transformer and circuit alike, necessitating protective elements that safeguard structure and person. Grounding systems with integrity and reliability are one such element, and circuit breakers and fuses both guard against overcurrents. The former offer convenience, while the latter feature reliability and quick response. All electrical protection devices are rated according to response speed, voltage rating, interruption rating, short-circuit current rating, and continuous current rating to afford electricians all the information they need to institute the best protective barriers. While commercial electricians should take all conventional precautions, like personal protective equipment, they should also adhere to the safety guidelines designed for commercial environments.

▶ Learning Objectives

- Compare commercial and residential protection methods
- Explain the function of a bimetallic strip
- Describe and apply different fuse classes and types
- Contrast short-circuit protection with overcurrent protection

▶ Key Terms

bimetallic strip
dual-element fuse
integrated fuse
interruption rating
limiter
magnetic element
overcurrent
overcurrent protection
short circuit
short-circuit current rating
short-circuit protection
thermal runaway
time-delay fuse
Type S fuse

The driving force behind electrical protection is, of course, safety. In addition to keeping customer needs at the forefront, electricians must develop a "sixth sense" for safety in all areas of the profession, from driving habits to and from sites to customer training on the maintenance and use of electrical service.

Hints and Tips
To maintain safety, always use protective equipment, keep current reference materials handy, and work "above code." Common sense should be the guide for completing installations and annotating project documentation. Something that appears unsafe probably is.

While residential and commercial installations differ in some ways, their protection methods and requirements are essentially the same. Residential applications feature 120V 20-ampere circuits and occasionally 220V circuits, as well as wire rarely exceeding 12 AWG. Commercial installations, in contrast, involve three-phase 240V, 480V, and 277V. One notable difference between the two installations is that ground fault circuit interrupter (GFCI) breakers may protect branch circuits in commercial environments.

▶ Understanding Circuit Basics

In any environment, when excess current travels a complete circuit or flows from a source through an external path, an **overcurrent** occurs. A **short circuit** is the ultimate overcurrent. With a short circuit, resistance is 0, affording, mathematically at least, infinite current flow.

Consider normal circuit operation in which a 120V source is connected to a 12-ohm load, creating a current of just under 10 amperes due to fuse, conductor, and termination resistances (Figure 7.1). Protection devices are $1\frac{1}{2}$ times the required amperage.

Now assume that a 0-ohm resistance ("shunt") to ground bypasses the 12-ohm load (Figure 7.2). Because

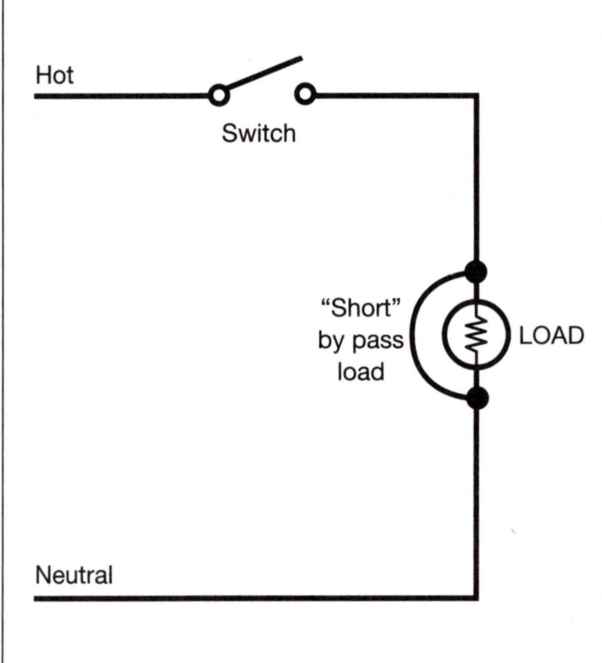

Figure 7.1 Short Circuit

ALERT!

Depending on such variables as temperature, environment, and line regulation, using part of the 50% buffer designed into a circuit may be hazardous. The 50% buffer is the difference between the expected current and the ampacity of the circuit. If, for example, the expected ampacity is 12 amperes, the ampacity of the circuit should be calculated as 12×1.5 or 18 amperes. Because there is no 18-ampere circuit, a 20-amp circuit would be needed. The 50% buffer is not designed for future expansion. Rather, it is designed for situations that occur randomly, such as excessive temperature or nonlinear loads.

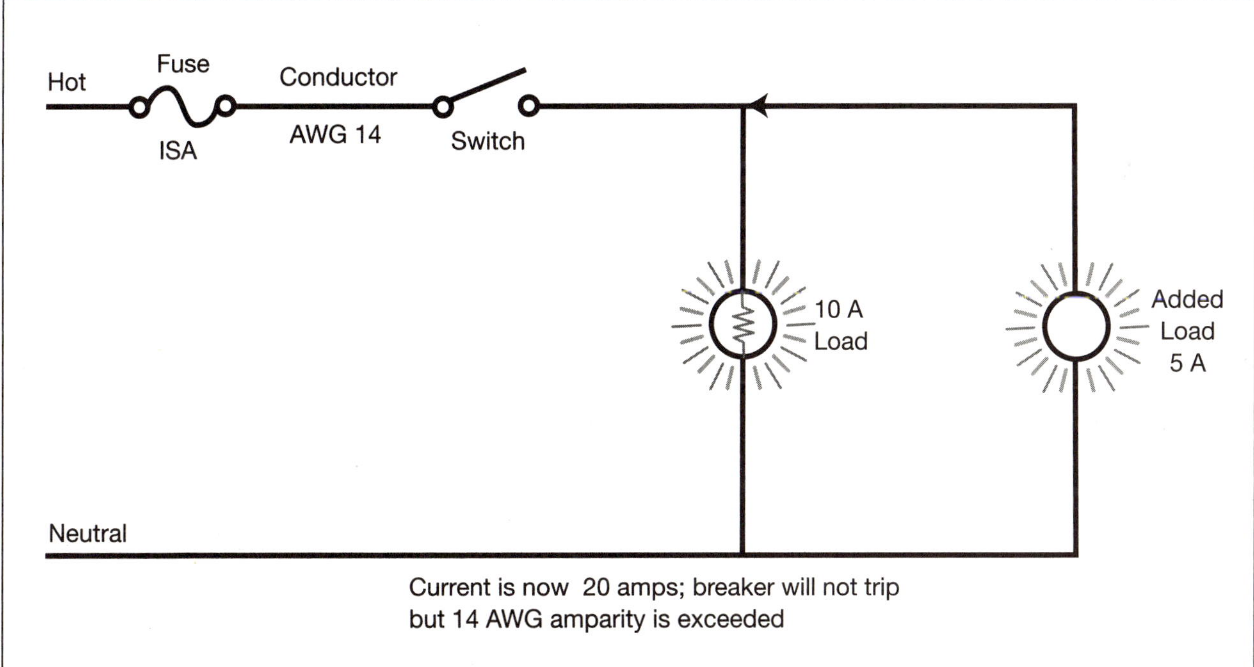

Figure 7.2 Overcurrent Condition

current follows the path of least resistance, the load has no current while theoretically, the shunt has infinite current because it is 0 ohms. Conductor and fuse resistance, possibly on the order of 0.001 ohms, determine short-circuit current. In this case, the short-circuit current could be on the order of 60,000 amperes. The fuse will blow, breaking the current, but if it does so too slowly the transformer can sustain damage.

The circuit simply has one too many loads to handle. When combined with conductor, fuse, and source resistance, the 6 ohms of resistance of the two 12-ohm paths create a current of just under 20 amperes. The extra load increases circuit current beyond the circuit's rating but under circuit capacity.

When a hot conductor contacts conduit, some assume no danger exists for two reasons: (1) because the current exits to ground through the metal conduit and (2) because the human body has more than 0 ohms of resistance. Due to improper joints, corrosion, or other factors, conduit's resistance may not be 0 ohms. In fact, with 60,000 amperes flowing to ground, it is likely that conduit could impart current to a body when touched.

When the human body has a resistance of 10,000 ohms, a reasonable average in dry conditions, and conduit has a resistance of 0.001 ohms, the body receives about 6 milliamperes of short-circuit current (0.001/10,000 × 60,000 amperes) which, even with normal conduit resistance, is dangerously close to the 10-milliampere lethal dose. Any factor that increases body current, such as a long conduit run, a corroded or deformed area, or wet feet, enhances the need for protection.

ALERT!

Even overcurrent insufficient to trip a breaker can prove hazardous under the right conditions. Overcurrent causes excess heat, which can lead to intermittent operation, failure, or even fire.

Did You Know?

The most significant protection steps include making sure a grounding system has integrity and reliability. This is done by verifying sufficient design ampacity and by ensuring solid connections per NEC Article 250.

▶ Instituting Short-Circuit Protection

All equipment is rated in terms of its ability to withstand a certain amount of current for a certain amount of time before opening or shorting, usually due to heat. In a standard automotive fuse, for example, excess current melts the fuse's wire, effectively breaking the circuit and thereby protecting the load.

Unfortunately, melting is not always the most advantageous means of **short-circuit protection**. In a transformer, primary and secondary windings around the core are typically in close proximity, separated only by thin conductor insulation that is designed primarily to promote magnetic energy transfer, not guard against shorts. When a transformer overheats, the insulation melts easily, shorting the primary and secondary windings together. When a circuit lacks short protection, the result is usually catastrophic for the transformer and problematic for the circuit.

Did You Know?

Short-circuit currents can exceed 50,000 amperes. Most manufacturers rate electrical equipment to withstand excessive currents.

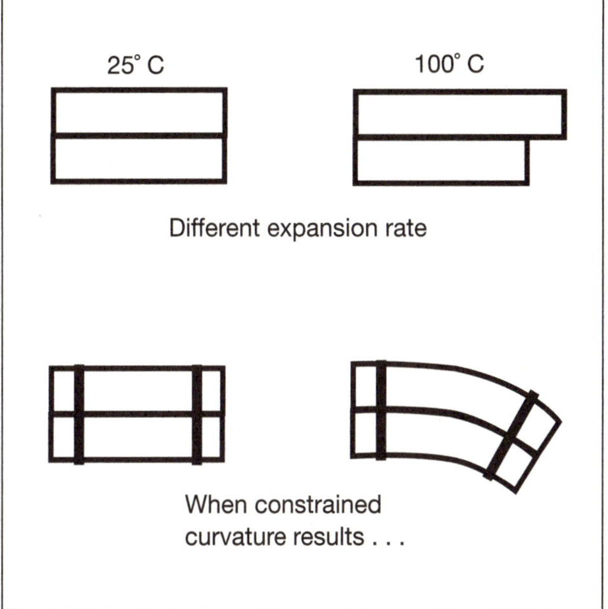

Figure 7.3 Bimetallic Strip

To guard against such undesired events, electrical devices have **short-circuit current ratings**, which are determined in part by Underwriters Laboratory (UL) test criteria and by the independent testing of such third parties as TUV. Short-circuit current ratings are either stamped on devices or specified in device literature. Each rating has two parts: the current the equipment can withstand and the time over which the equipment can withstand that amount of current. For example, a device may allow 200 amperes for 300 milliseconds.

Temperature is the main concern with excessive currents, especially short circuits. Electrical protection is designed to accommodate the temperature limitations of conductor insulation. When conductor insulation becomes compromised at 97° C, for example, protection limits current to a value that keeps conductors below this temperature. To track temperature changes, most motor contactors and controllers use devices like **bimetallic strips**, which change shape according to temperature (Figure 7.3).

ALERT!

To ensure proper functioning, a piece of equipment's current rating must align with the ratings of its protective devices. If, for example, a device can withstand 100 amperes for 2 seconds but its protective devices allow 100 amperes for 3 seconds, the device will fail before its protective devices, creating a possible hazard.

Did You Know?

The relationship between current and temperature is not linear. Doubling current does not double temperature. Instead, the square rule applies. Doubling current, for example, increases temperature by a factor of four (2 × 2). Tripling current increases the amount of heat generated nearly tenfold (3 × 3).

Did You Know?

A current **limiter** has two or three protective elements: (1) a thermal element, which protects in the event of slight but significant current overloads; (2) an **integrated fuse**, which guards against high short-circuit currents; and (3) a **magnetic element**, which is used to protect against low short-circuit currents.

At high temperatures, conductors lessen their resistance, thereby passing more current. In a concept known as **thermal runaway**, the added current increases temperature, which increases current. In such circumstances, two options are available: (1) limit the current or (2) do not limit the current. Some protective devices illustrate the first option by limiting the current passing through them. Instead of tripping, they reduce circuit current by increasing resistance. If a 12-ohm load were to fail, for example, an effective protective device would impart a load of up to 12 ohms to restrict the current. Such protective devices ease the burden of selecting conductors that can withstand a short-circuit current for the specified time.

Hints and Tips
The response speed of an electrical protective device appears on the manufacturer's data sheet.

Conductors that meet short-current requirements are needed when the second option, the option of not limiting current, is exercised. Appliances that need added current at startup, such as motors and dynamos, might need such conductors. For added detail on these requirements, see NEC Tables 310.16 and 250.122.

▶ Instituting Overcurrent Protection

Overcurrent protection takes two forms: circuit breaker and fuse. As mentioned earlier, circuit breakers are the more convenient option, while fuses are generally quicker and more reliable. In residential installations, overcurrent protection is limited to 15- and 20-ampere breakers. In 240V circuits, dual-pole breakers are sometimes used.

▶ Rating Electrical Protection Devices

Electrical protection devices, which are rated according to response speed, voltage rating, **interruption rating**, short-circuit current rating, and continuous current rating, are available in standard values from 1 to 6,000 amperes. Below 800 amperes, intermediate values are rounded up to the next higher standard rating; above 800 amperes they are rounded down to the next lowest standard rating (NEC Article 240).

Assessing Speed of Response

In a protective device, response speed is the time from onset of an overcurrent condition until the device breaks the circuit. This variable depends on how quickly the sensing device heats to melt, as in a fuse, or pulls out of the way of a spring-loaded switch, as in a breaker. Some sophisticated equipment uses a zero-cross detector to turn the supply off when it is 0, which happens twice in every cycle of a sinusoidal input wave, or about every 8 milliseconds. While this condition never occurs in three-phase systems, phases can be shut down individually at about every 2.33 milliseconds.

Identifying Voltage Rating

Because circuit capacity must at most equal protective device capacity (NEC Section 110), a 120V branch circuit requires a protective device rated 120V or higher. In 120V, 208V, or 240V systems, 600V devices designed for 480V circuits are commonly used. Typically, breakers are rated significantly higher than their circuits.

ALERT!

Circuit breakers rated below circuit voltage jeopardize system reliability. The increased voltage could cause an arc condition, which represents a significant fire hazard.

Working with Interruption Rating

In electrical protection, interruption rating is the amount of current a device can reliably stop. Motorized toys use little current, while automobiles use a great deal.

Defining Short-Circuit Current Rating

A device's structural integrity under an arc condition is known as its short-circuit current rating. Device ratings must exceed the current available in circuits. As mentioned earlier, the resistance of terminations, conductors, and inline devices, as well as supply-line limitations, prevent short-circuit current from reaching infinity.

Discussing Continuous-Current Rating

Short-circuit currents are measured in terms of duration and intensity. For continuous-current ratings, duration is assumed to be infinite. Continuous-current rating limits intensity. Assume, for example, that a switch serving a receptacle is expected to allow 10 amperes through a circuit. Continuous-current rating will limit intensity to a manageable level, although the switch will likely allow more than 10 amperes through.

Hints and Tips

To ensure a protective device can handle a targeted current draw, consult the manufacturer's data sheet for the device's continuous-current rating.

Hints and Tips
When replacing fuses, be sure to match the original's form, fit, and function.

▶ Classifying Fuses

Although fuses have, for the most part, given way to circuit breakers due to convenience, they are still used when quick response or increased reliability are required. Fuses come in three types. The first type, standard link, is the fast-blow type. Such fuses are unsuited for motor use. The second type, delay link or loaded link, imparts a slight delay and is useful for small motors and other surge devices. The third and final type, the **dual-element fuse**, has short-circuit or fast-acting elements in series with slower-acting links that fail at less-than-short-circuit currents. These fuses are ideal for motor control and protection.

Per NEC, fuses are classified based on amp and voltage ratings, as well as the ability to limit current. Boxes that hold current-limit fuses may not accept fuses that do not limit current, because branch circuit specifications may vary when current limiting is used (NEC 240.6). Class G fuses are available in less than 1 amp up to around 60 amps (Figure 7.4). They are small, current-limiting fuses that are available at voltage ratings of up to 600VAC. Interruption rating is around 100,000 amps.

- Non-renewable
- 12 second time delay @ 200%
- Current limiting
- 600 VAC (0.204) and 480 VAC (25–60A)
- Different sizes for different current ratings

Figure 7.4 Class G Fuse

Class H fuses are standard low-cost, non-current-limiting fuses that are available in currents up to 600 amps and at voltage ratings of up to 600V (Figure 7.5). Class H fuses can interrupt currents of up to 10,000 amperes, which is fairly low. Class J fuses are current-limiting fuses with interruption ratings in the 200,000-amp range (Figure 7.6). Available in fast-acting and

- General purpose branch circuits
- Lighting circuit
- Not for use with inductive loads
- Renewable and non-renewable
- Not avaiable for time delay applications
- Available with DC ratings

Figure 7.5 Class H Fuse

Did You Know?

Like circuit breakers, fuses are housed in boxes. From the outside, breakers and fuses look identical.

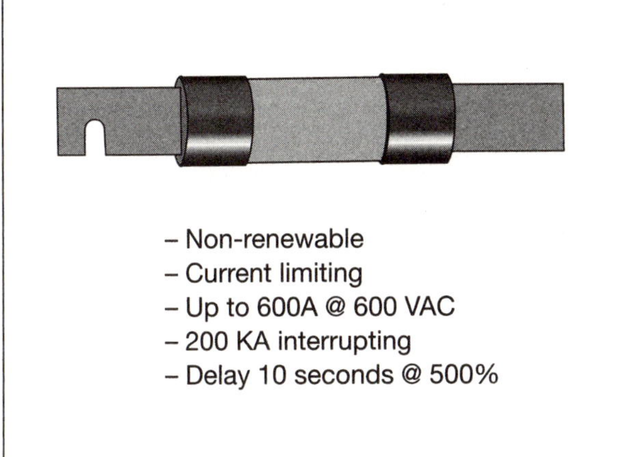

Figure 7.6 Class J Fuse

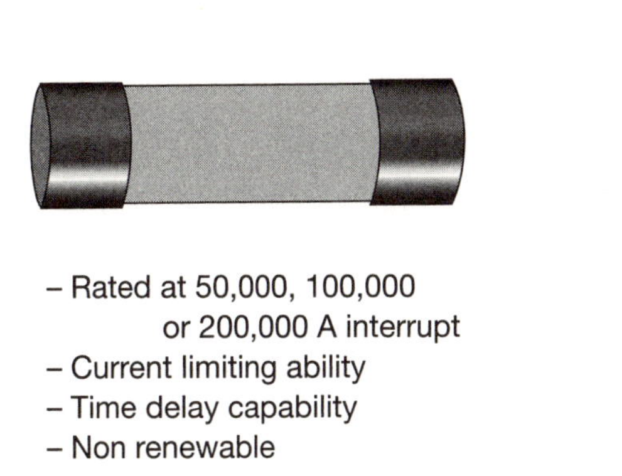

Figure 7.7 Class K Fuse

time-delay versions, at 600V these fuses range from a few amperes to 600 amperes. Class K fuses are non-limiting fuses with interrupting ratings in the 50,000 to 200,000 ampere range (Figure 7.7). These are **time-delay fuses**, with delay times of around 8 to 10 seconds. Available from under 1 ampere to 600 amperes, Class K fuses are rated at up to 600VAC. Class L fuses are designed for high currents, those from 600 to 6,000 amperes (Figure 7.8). Available as time-delay or fast-action, these current-limiting fuses have an interruption rating in the 200,000-amp range.

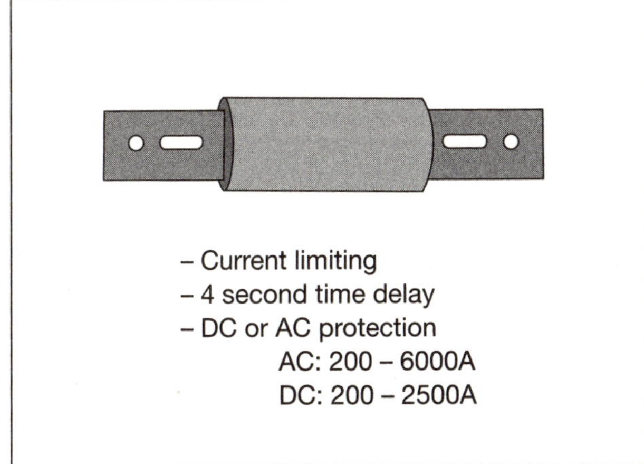

Figure 7.8 Class L Fuse

The practice of improper fuse substitution inspired development of the **Type S fuse**, which comes in 0–15, 16–20, and 21–30 amp variations (Figure 7.9). The Type S fuse supersedes the Edison fuse found in old residences (Figure 7.10). While the Edison fuse could accept fuses of any amp rating, each Type S has a unique thread, making it impossible to exchange for another. Edison fuses are designed for use only in 120V applications, including those with grounded neutrals that allow 240V with two separate hots. The female side, or screw flange, is connected to ground; the center button is connected to hot.

Hints and Tips

To test fuses on live circuits, attach the red or hot lead of a voltmeter to the hot side of the fuse and the ground lead to the ground side. When a fuse is intact, voltage should be nearly 0. When power to the circuit is shut down upstream, remove the fuse and check it with an ohmmeter. It should read close to 0 ohms.

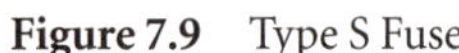

Figure 7.9 Type S Fuse

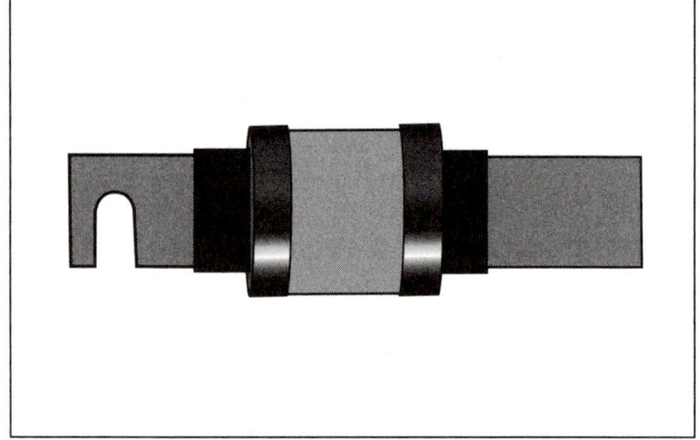

Figure 7.10 Edison Fuse

Job Connection

As his final project, Dustin set out to change 12 lamps high in his trade school's atrium. In preparation, he took all the proper steps: secured building blueprints from his instructor, ordered spare lightbulbs, cleared a path for the rented scissor lift, laid down tarps and cardboard to keep the floors clean, and established boundaries for students and employees in the building.

Dustin climbed the lift with the replacement bulbs in hand, as well as a box for the discarded ones. As he removed the bulb from the first fixture, he realized that the two bulbs differed. Confused, he confirmed that the bulbs he had purchased were those specified on the prints.

To rectify the problem, Dustin assembled all evidence, including the old and new bulbs and the school's blueprints, and set out to develop a plan for his customer's satisfaction. Dustin was nonaccusatory when he approached his customer. "These are the drawings I was given with the lamp specs," Dustin started, "and here are the bulbs I purchased. As you can see, my bulbs match the prints, but I found different bulbs in the fixtures." Dustin then asked his customer for a potential solution, already prepared with a solution of his own.

Once Dustin and his customer agreed on a solution, he set out to execute the plan. Next time, though, Dustin resolved to do things a little differently. For example, he would physically inspect any bulbs before ordering replacements. He would also ask for the most recent service ticket on any apparatus to ensure he had the most current information. Finally, he would likely rent a scissor lift again, but not before confirming he had the proper replacements.

▶ Working with Circuit Breakers

Circuit breakers operate on the principles of heat expansion and contraction. Heat causes one side of a breaker's bimetallic strip to expand faster than the other. As a result, the strip bends. As the strip bends, it slides out of a notch in a spring-loaded plunger. The spring presses the plunger forward and breaks the circuit. Once the strip cools and assumes its original shape, the lever can slide the plunger back against the spring into its notch, and the breaker can assume normal function.

NEC Section 240 outlines a number of circuit breaker requirements. First, a breaker's manual control (switch) must be clearly marked "On" and "Off." Also, breakers must be derated by 20% for such constant load requirements as office lighting or computer servers. Breakers must be certified for switchgear use and marked as such, just as they must be marked HID, for high-intensity duty, or SWD, for switching duty, when used as switches. When breakers are installed in panels or boxes, their ampere ratings must be stamped and visible. Finally, breakers must be able to trip when in the "On" position so that circuits can be broken when switches are taped in the "On" position.

▶ Installing HVAC Applications

Special rules govern heating, ventilating, and air conditioning (HVAC) applications. It is important, for example, to maintain a disconnect within line-of-sight of HVAC equipment to promote safety. With such a disconnect, people distant from the system, for example, cannot inadvertently switch the system on because they are hot/cold. HVAC equipment must be protected at a disconnect according to information on its nameplate.

> **Hints and Tips**
> To ensure proper HVAC installations, simply meet or exceed the requirements spelled out on equipment nameplates.

▶ Delivering Personal Safety and Protection

As in residential installations, the electrician's personal protection in commercial environments is paramount. While commercial electricians should take all the precautions that apply for residences, including using insulated tools and donning personal protective equipment, they should follow the safety guidelines specific to commercial environments.

One such guideline is the lockout-tagout procedure mandated by the Occupational Safety and Health Administration (OSHA) for motors above 240V. The lockout-tagout procedure requires an electrician working on a piece of a system to physically disconnect the

> **Hints and Tips**
> To be prepared for commercial installations, keep a set of lockout-tagout equipment in the vehicle.

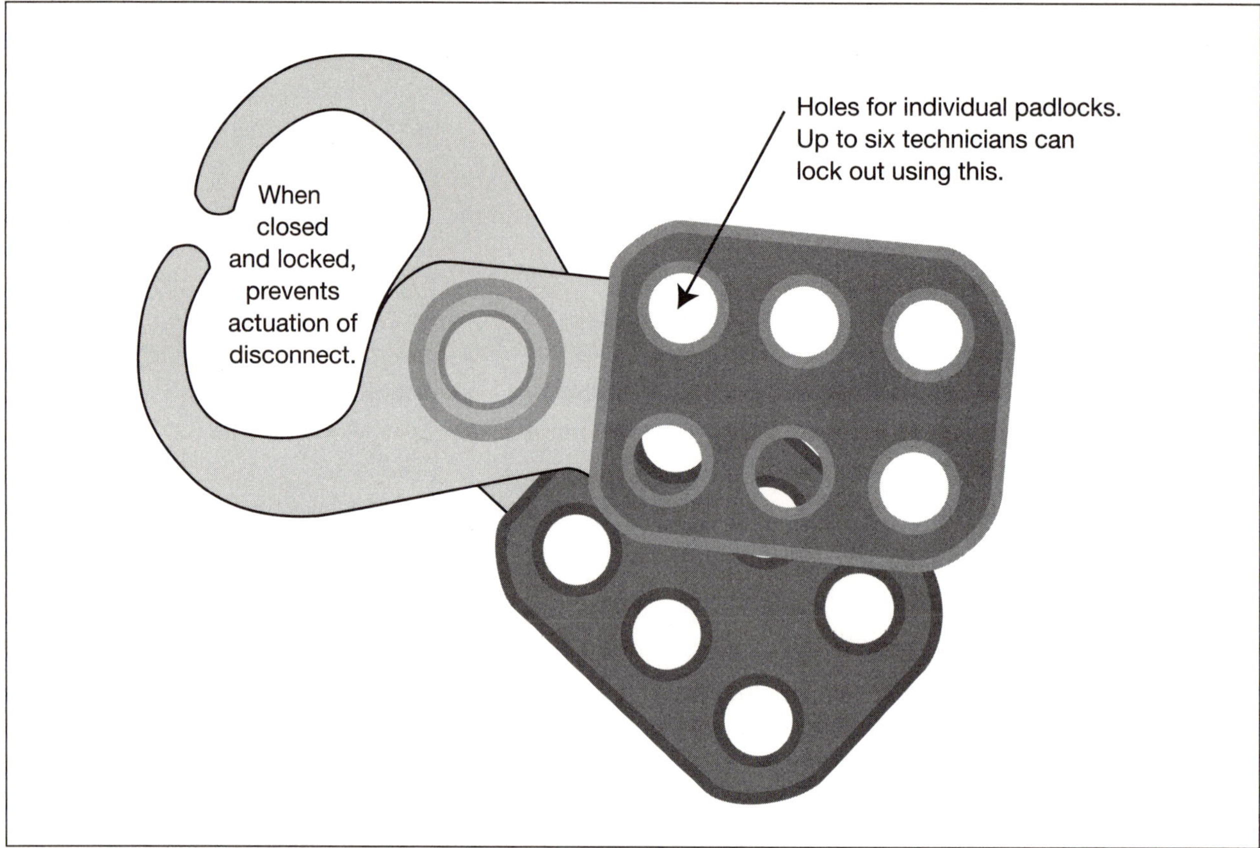

Figure 7.11 Lockout-Tagout Gear

system's power supply and then install padlocks to keep the system off while completing work. When several electricians work on a project, the first to arrive removes power at the disconnect and secures a personal lock to the equipment via a holed device (Figure 7.11). As other electricians arrive, they secure their personal locks in similar fashion. As they finish, they remove their locks. The system is reenergized only when all electricians agree that doing so is safe.

▶Chapter Review

1. How do standard breakers and GFCI breakers differ in appearance?

____

____

____

____

2. Installation of raceways under roof decking must be ____________ below the decking.

3. Which of the following is *not* a standard amperage value?

a. 100
b. 400
c. 600
d. 750

4. Panelboards are limited to 42 protective devices.

a. true
b. false

5. An electric water fountain in a public building requires GFCI protection.

a. true
b. false

6. Electric pipe or plumbing heaters (wrap-arounds) require signage

a. every 10 ft.
b. every 20 ft.
c. adjacent to supplied or supplying equipment.
d. b and c

7. A relay may be used as part of an overcurrent protection scheme.

a. true
b. false

8. It is permissible to use a timer to control motor operation and disconnect motors for servicing.

a. true
b. false

9. Motors operated at greater than ________ volts must have lockout-tagout systems.

a. 120
b. 240
c. 480
d. 600

10. What is Greenfield?

____

____

____

____

11. Electricians should carry portable lockout-tagout sets for high-voltage locations.

a. true
b. false

12. Local codes, not NEC guidelines, are the top authority for connecting to an electrical supply.

a. true
b. false

13. In a continuous load, rated current is expected to flow for at least ________ hours.
- **a.** 2
- **b.** 3
- **c.** 12
- **d.** 24

14. When wiring a three-phase panelboard, how must phases be arranged?

____

____

____

____

15. NEC Chapter 5, which covers hazardous locations, ______________ NEC Chapters 1–4.
- **a.** supersedes
- **b.** supplements
- **c.** modifies
- **d.** repeats

16. Vending machines require GFCI protection.
- **a.** true
- **b.** false

17. All metallic boxes and components must be grounded, unless a grounding conductor is not used.
- **a.** true
- **b.** false

18. Per NEC ______________, a panelboard requires a circuit directory.

19. Calculations for the protection of motor circuits and nonmotor circuits are the same.
- **a.** true
- **b.** false

20. Leaving unused panelboard spaces open to improve the ventilation in hot or damp areas is good practice.
- **a.** true
- **b.** false

21. The requirements for burying 240V conductors appear in NEC ______________.

22. The maximum motor horsepower that can be disconnected using the panel breaker is:
- **a.** 0.1
- **b.** 0.125
- **c.** 0.25
- **d.** 1.0

23. Installation of a floor drain in a vault intended for wet transformer(s) is
- **a.** always required.
- **b.** required when the vault has more than 100KW of transformer capacity.
- **c.** good practice regardless of rules or requirements.
- **d.** none of the above

24. Tightening a bolt or fastener too much violates NEC regulations.
- **a.** true
- **b.** false

25. How would demand be calculated in a commercial environment where a branch circuit could effectively handle 20 KW?

____

____

____

____

26. Direct current motors are always protected at a minimum of 150% of full load current.

a. true

b. false

27. What does the phrase *in sight* mean?

____

____

____

____

28. NEC requires that circuit breakers rated at ________ amperes or fewer and ________ volts or fewer have their ampere ratings molded, stamped, etched, or similarly marked into their handles or escutcheon areas.

a. 75, 300

b. 100, 600

c. 80, 300

d. 90, 600

29. Receptacles as specified in NEC Article 647 may not be used for which of the following?

a. motors

b. lighting

c. telecom equipment

d. computers

30. Splices and taps in a wireway must be inaccessible to ensure safety.

a. true

b. false

PART

III

At the Plant: Industrial Applications

CHAPTER

Industrial Application

CHAPTER SUMMARY

The in-house electricians charged with planning, executing, and maintaining commercial electrical installations, conversions, and upgrades must understand, and be able to implement, the energy-savings and control functions increasingly gaining prominence in the industry. Armed with a solid understanding of electrical basics, such as efficient power use and equipment sequencing, commercial electricians can both execute the needed control programming and tailor large-scale control systems to specific needs. As business continues to progress and electricians delve ever deeper into realms traditionally held by other technical professionals, it will be ever more important for electricians to be able to use electricity and electrical prowess to harness the automated technology that will be driving contemporary security, access, and other daily activities.

Learning Objectives

- Explain the system of jurisdiction over industrial installations
- Discuss basic electrical principles and their applications
- Outline programmable logic control (PLC) as an industrial tool
- Identify the basic types of electrical control systems and how they are used
- Describe the system for classifying electrical control devices
- Give the general principles of control programming

Key Terms

capacitor
inductor
ladder diagram
optimum power level control (OPLC)
power factor logic
programmable logic controller (PLC)
resonant frequency
rung
zero point data transfer

In practice, the NEC maintains no jurisdiction over most industrial installations. Instead, most industrial entities like factories and plants employ in-house electricians to work with local authorities to plan, execute, and maintain any electrical installations, conversions, or upgrades. Local building and/or zoning commissions set governing rules and policies in this realm, in concert with organizations like the National Fire Protection Agency (NFPA), which publishes the NEC.

> **Did You Know?**
>
> Given the scope and complexity of their work, industrial electricians often must complete advanced degrees.

Discussing Electrical Basics

Efficient power use is a driving goal of industry. Many communities base industrial electrical rates on peak usage, defined as maximum power use over a short period. If, for example, a plant's electrical demand is at 100,000W in the morning when machinery starts but 35,000W in the afternoon when production diminishes, the plant's electric rate will be based on the 100,000W morning peak.

To achieve more favorable rates, plants can do things like sequence in machines at startup, which lessens demand and therefore the basis of electrical rates. To achieve sequencing, industry uses such tools as discrete timers, relays, and **programmable logic controllers (PLCs)**, which are discussed later. A basic understanding of electrical concepts helps support a general understanding of sequencing technologies and benefits electrical contractors and tradespeople alike.

Delving into Energy Expenditure

To illustrate the basic concept of energy expenditure, assume you are running a large, motorized saw. Given that wood cutting is the goal, the energy needed to

Did You Know?

Energy savings and control functions are becoming increasingly important in residential applications as well as commercial ones. In "smart homes," for example, small PLCs are doing things like turning off lights in unoccupied rooms.

Hints and Tips
In typical electrical circuits, natural frequency is 60Hz.

bring the saw up to running speed is wasted, because no wood can be cut until that point. Similarly, the energy needed to keep the saw running once cutting is completed is wasted, because cutting must cease before the saw turns off. Overall, the only truly useful energy is that expended during cutting.

The "power factor" calculates the amount of useful energy as follows:

$$\text{Power factor} = \frac{\text{(resistive power)}}{\text{(total power)}} \times 100\%$$

A power factor of 95%, for example, translates to 95% use of electricity.

The **power factor logic** of a piece of equipment, or a whole factory, can be improved by applying an equal and opposite reactance to the source. The two factors cancel each other out in a form of resonance and use no energy. A pendulum demonstrates this principle.

Illustrating Real-World Energy Concepts

At its core, a pendulum is an energy-transfer device. At rest, it has no energy. Once in motion, however, it gains energy, trading potential energy, or height, for kinetic energy, or velocity. A rising pendulum increases its potential energy by increasing its height, yet decreases its kinetic energy by decreasing its speed as it rises. At its maximum height, a pendulum has its highest potential energy and zero kinetic energy, or zero velocity. As it swings back downward and reaches its vertical midpoint, its speed maximizes, which maximizes kinetic energy. Potential energy, at this point, drops to zero. A pendulum continues swinging until friction slows its speed and decreases its kinetic energy, converting all available energy to heat.

Swing rate is directly proportional to **resonant frequency**. A slowly swinging pendulum, for example, has low resonant frequency. The weight of a pendulum and the length of its string determine pendulum swing rate. Increasing pendulum weight or string length decreases swing speed and therefore lowers resonant frequency. Conversely, decreasing pendulum weight or string length increases resonant frequency by increasing swing speed.

Whereas a pendulum trades kinetic and potential energies, an electrical circuit trades magnetic and electrical energies. Further, pendulum and swing translate to inductance and capacitance. To cancel each other

Did You Know?

Wires get hot because electrons create friction between themselves and other particles as they travel through a conductor.

out and resonate at line frequency, capacitance and inductance must have the same yet opposing response to power. A **capacitor** acts like a small battery, storing voltage. A fully charged capacitor has maximal potential energy. A capacitor discharges into an **inductor**, decreasing its potential energy yet increasing its kinetic energy. Discharge builds a magnetic field as current through the inductor increases. Once current through the inductor ceases, the magnetic field collapses. The result is current, which charges the capacitor for another cycle. This cycle repeats until friction from wires (resistance) ends the capacitor/inductor energy trade-off. Machines with low power factors have small differences between capacitive and inductive effects.

▶ Studying Electrical Control Systems

The electrical control systems that help render favorable electrical rates execute a sequence of events based on one of two runglike approaches: basic design, as is the case with discrete ladder logic, or programming, which is the case with PLC control systems. Discrete ladder logic or relay/timer technology uses limit switches, sensors, and timers to open and close contacts and energize various functions. A control system for a conveyor belt that moves in two directions is one example (Figure 8.1). Using normally closed and normally open contacts, button presses at either end of the

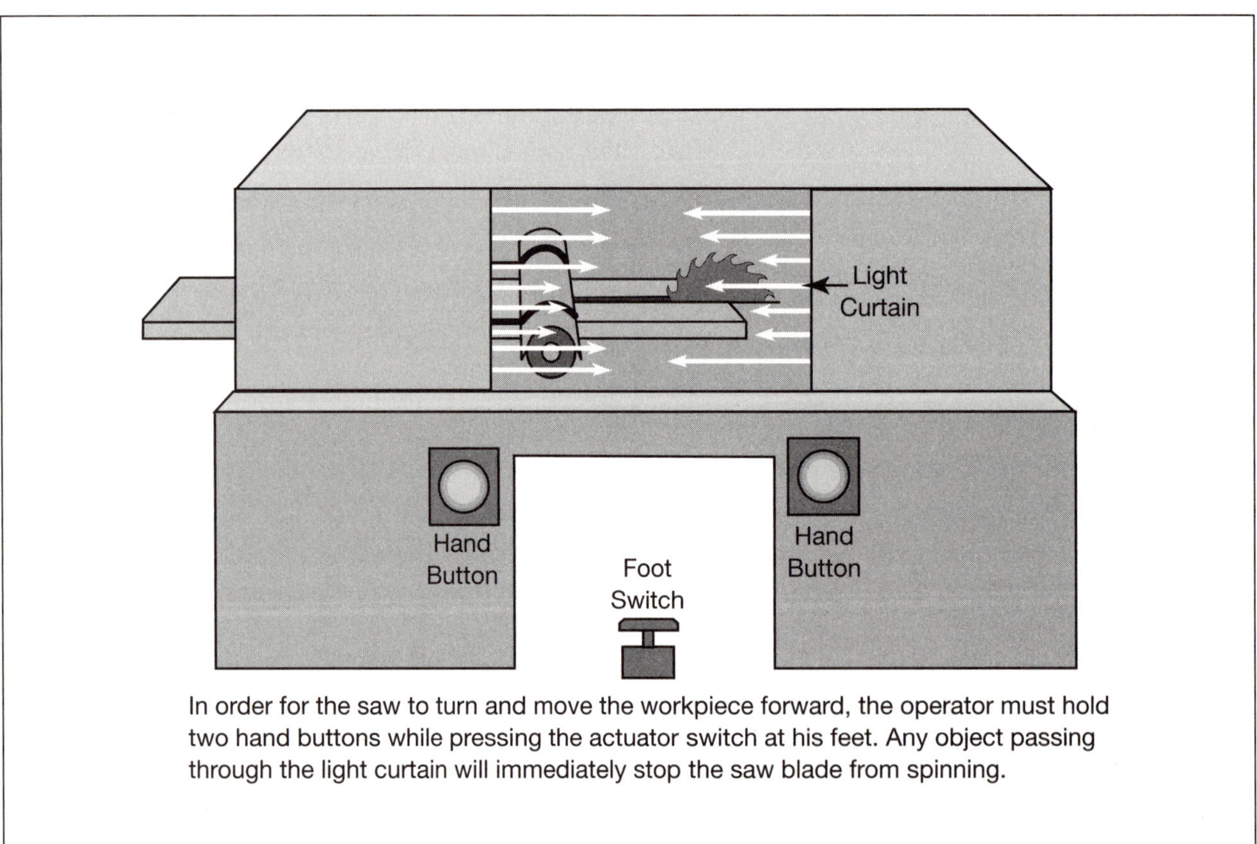

In order for the saw to turn and move the workpiece forward, the operator must hold two hand buttons while pressing the actuator switch at his feet. Any object passing through the light curtain will immediately stop the saw blade from spinning.

Figure 8.1 Relay Ladder-Logic System

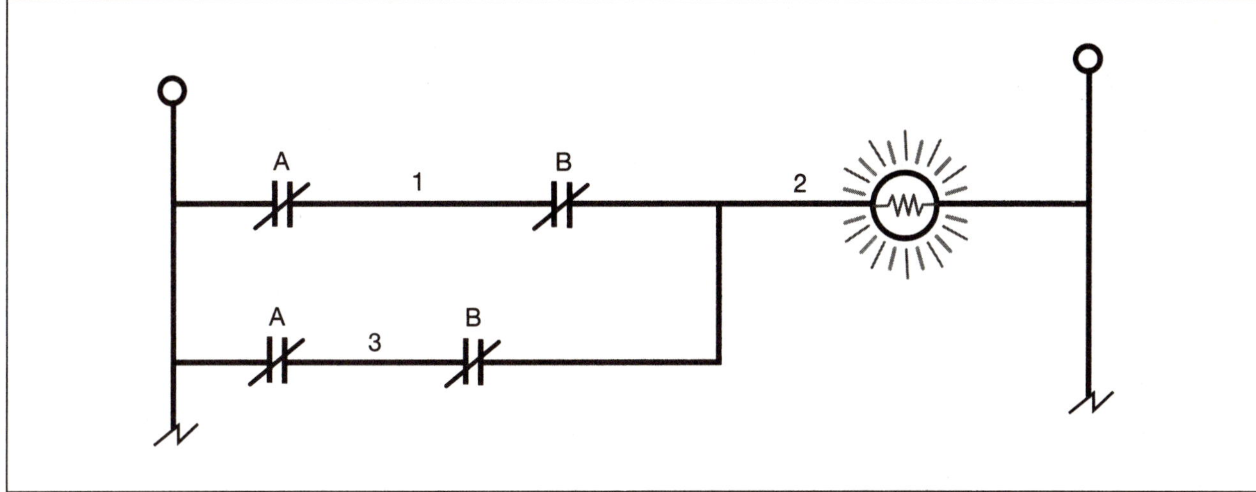

Figure 8.2 Relay Ladder-Logic Diagram

belt execute the desired actions. Circuit design executes the decision-making process in a **ladder diagram** (Figure 8.2).

Like discrete ladder logic, PLC control can facilitate decision making. In a sample PLC control system (Figure 8.3), switches route to inputs and outputs turn a motor forward, backward, or off. Low-voltage signals control high-voltage relays as well as all PLC circuits (Figure 8.4).

A security system is a specialized PLC for home and commercial use that uses low-voltage (5–24V) signaling devices. These devices usually interface with a

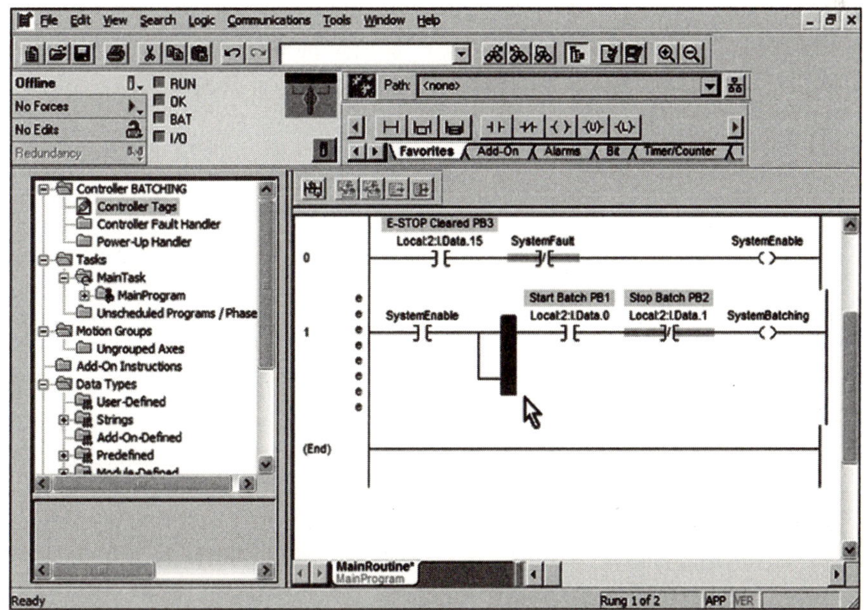

Figure 8.3 PLC Control Program

Figure 8.4 PLC Control Circuit

building's power grid to control emergency lighting, open and close doors, and energize sprinkler systems. Many security systems also work with telecommunication installations to make telephone calls.

Like security systems, HVAC systems are often controlled by PLC technology. One Caribbean resort features a PLC controller that uses automated switching to "remember" a room's HVAC and lighting environment, or "snapshot." When an occupant exits a room, the controller does things like turn off lights and air conditioning. When that occupant returns and engages the room's keycard, the controller reinstates the room's lighting and temperature levels.

Did You Know?

Based on its cost-effectiveness and ease of use, **optimum power level control (OPLC)** technology is gaining acceptance in homes and businesses. In heating, ventilating, and air conditioning (HVAC) systems, for example, variable-speed drives are replacing constant-speed fans and motors, eliminating the need to constantly start and stop motors, which consumes energy. While the relative simplicity and efficiency of OPLC systems is earning them growing popularity, they are initially more expensive and require training.

Tailoring Control Systems to Needs

To meet unique customer needs and accommodate changing requirements, electrical control systems sometimes need tailoring. Assume, for example, that an industrial electrician must retrofit a conveyor circuit with a safety feature that will move the conveyor only when operators on both sides are depressing buttons. A discrete logic circuit for this system would require an almost complete redesign. A similar PLC system, in contrast, would require only a few added lines of code. In addition to easing the retrofit, the PLC circuit would allow production to continue with far fewer interruptions.

Whatever the circuit type, the best approach to connecting the two switches in this safety feature would be to wire in series. That way, disengaging one button would interrupt the entire circuit. In a PLC circuit, something called the AND logical function achieves this goal. In Figure 8.4, the FWD-IN button connects to Input 0, and FWD-OUT connects to Input 1. The programming statement "IF Input0 =1 AND Input1 =1 THEN motor CCW" translates to, "If the inside forward button and the outside forward button are pressed at once, turn the motor counterclockwise to move the conveyor forward."

Using only two values, 0 and 1, digital logic converts inputs to outputs. Every number, letter, and value has a digital equivalent. The digital equivalent of the number 9, for example, is 1001. Because 0 and 1 are the only digital values, the only questions a system can answer are "yes" and "no," such as "Is the front door open?" or "Is the temperature less than the value stored in memory?"

Classifying Electrical Control Devices

Using digital 0s and 1s to answer yes-or-no questions, a PLC or another controller examines system inputs (e.g., temperature), makes a decision (e.g., decrease temperature), and signals the input condition (e.g., turn on cooling). Usually, low-voltage devices provide controllers' determining information.

Reed or limit switches, as well as other sensors, indicate when boxes reach a certain point on a conveyor or doors or windows are open. Optical switches, for their part, can signal when occupants enter or exit buildings, when mail is placed in mailboxes, or when paper is placed in copy machines. At night, these sensors can engage inside or outside lighting. Thermocouples provide temperature-dependent resistance. By converting voltages to digital bit patterns, these controllers indicate when some action is needed. Pressure sensors, strain gauges, humidity transducers, and pH transducers that measure alkalinity or acidity are yet more device examples. Automobiles illustrate how sensors or transducers transmit physical quantities to controllers. Sensors read the outside world statically, while transducers read motion and convert it to an electrical signal (Figure 8.5).

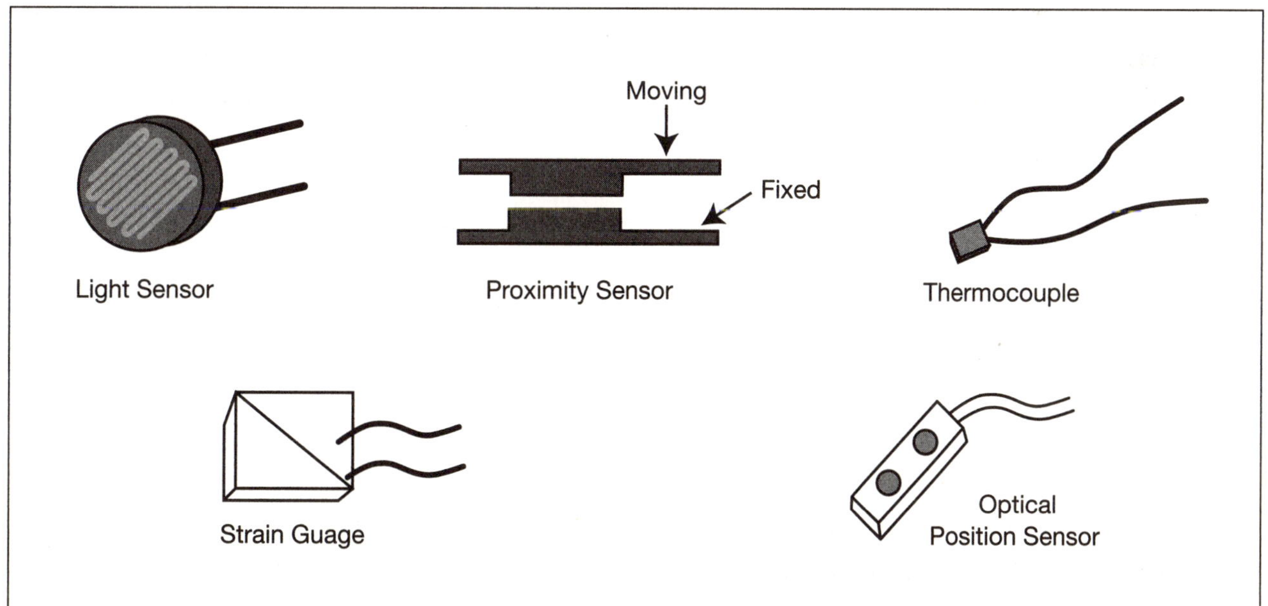

Figure 8.5 Sensors and Transducers

In addition to facilitating yes-or-no states, control logic executes timing schemes. Consider a garage door opener, for example. With a command from the opener's remote as the input, the opener will respond by opening the door and waiting 30 seconds. If a vehicle fails to enter the garage within 30 seconds, signaled by a broken beam across the door, the opener will close the door. A security system demonstrates a more intensive timing scheme. The following program format shows a sample "to do" list for security hardware:

Input: ARM button for door and window sensors on the keypad.
Response: Wait 2 minutes.
Is Door 1 open? If YES, sound alarm.
Is Door 2 open? If YES, sound alarm.
Is Window 1 open? If YES, sound alarm.
Is Window 2 open? If YES, sound alarm.
Has the RESET key been pressed? If YES, turn off alarm, disarm the system, and go to WAIT.
Go to Door 1.
Go to WAIT.

Once the security system is armed, it completes all steps in the sequence, checking the switches on all windows and doors. When a window or door is closed, the system closes its switch, completing the circuit. The program then executes the next step, and checks Door 2. If Door 2's input has no voltage from an incomplete or a broken circuit, for example, the system responds by initiating an alarm. The reset switch directs the system to turn off the alarm and take the system off-line.

Executing Control Programming

A program, then, is a series of steps carried out by hardware. Electronic control devices are typically programmed by their manufacturers or through programming software. Programmers can change, delete, upload, or download programs in seconds. PLCs can execute those programs in a few milliseconds. A statement that looks for a reset and turns off an alarm, for example, may occur 20 times in 1 second. There is no "control lag."

Many PLC programs can be developed off-line using **rungs** or ladder logic, tested through simulation, downloaded to controllers, and then verified, all within a few moments. Immediacy is essential in industrial environments, because results are tied to production and throughput.

▶ Looking to the Future

While automated technology today is found mainly in industrial or manufacturing environments, in the future it will find its way into homes and businesses as a means of managing security, access, and other day-to-day activities. In demonstration of such "smart technology," refrigerators will keep running inventories of foodstuffs, e-mail grocers for supplies, and pay bills, while computer interfaces will facilitate the monitoring and control of a home's electrical loads. Residents kept late at work will be able to e-mail their stoves from their cell phones or geographic positioning systems (GPSs) to ensure their stoves prepare dinner promptly.

Already, small-scale, affordable PLCs in the $100 range handle most tasks required outside manufacturing environments, as well as some within. High-end models are IP addressable, meaning PLCs or the Internet can control them remotely. As the world advances, HVAC and its control will focus on enhanced efficiency to mitigate fuel costs and improve the currently widely used single-zone heating. Also on the horizon is zero point data transfer, technology by which small chunks

Did You Know?

Unlike PLCs, ladder logic circuits require complete redesigns whenever changes arise.

of digital data are sent over power lines during those brief but repetitive periods when sine waves cross 0 (Figure 8.6). **Zero point data transfer** is like having a conversation while someone is sawing wood nearby: Only short, quick interchanges are possible between saw cuts. If you could speak and listen fast enough, though, this mode of speech would take no longer than conventional conversation.

All such futuristic scenarios are well within reach, and even common in contemporary high-end homes and commercial buildings. Electricians, for their part, are an integral cog in technological implementation. As control logic takes hold, for example, electricians will have to learn to use and troubleshoot it. If zero point data transfer edges out fiber optics, electricians will have to consider the quality of waveform that is delivered, as well as the quantity and distribution of power.

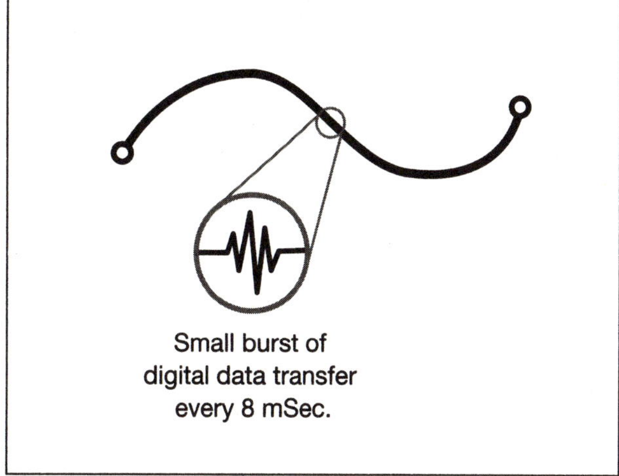

Figure 8.6 Zero-Point Data Transfer

Clearly, the scope of electricity is expanding. Some day, electricians may well assume tasks traditionally held by HVAC technicians, security system installers, and Internet/computer technicians.

▶Chapter Review

1. Give three examples of household items that use PLC technology.

____

____

____

____

2. A relay senses small currents to control larger ones.

a. true

b. false

3. Which of the following has primary jurisdiction over industrial electrical installations?

a. local agency

b. NFPA

c. NEC

d. all of the above

4. Digital devices and systems use 1s and 0s.

a. true

b. false

5. A thermometer or thermocouple is a sensor.

a. true

b. false

6. The time it takes for a PLC to take action once the appropriate signal is received is called __________.

7. A device that receives signals, makes decisions, and generates control signals based on its inputs is a

a. PLC.

b. computer.

c. CPU.

d. all of the above

8. When a system's total power is 10KVA, 6KVA resistive, the power factor is

a. 16%.

b. 32%.

c. 40%.

d. 60%.

9. If a system in one quadrant of a commercial operation, Q7, has a low power factor due to excess capacitance, the electrician should

a. remove the capacitor.

b. use thicker wire.

c. add an inductor.

d. turn off the unit.

10. Most plants and factories contract with local authorities to plan and execute their electrical installations.

a. true

b. false

11. Peak demand usually occurs in

a. overnight hours.

b. early morning.

c. late morning.

d. afternoon.

12. The time a PLC needs to complete an iteration of its program is ____________.

13. A PLC that is programmable over the Internet or through a network must be which of the following?

a. IP addressable
b. equipped with a CD drive
c. developed
d. all of the above

14. A normally open switch has an inner spring to keep the switch deactivated.

a. true
b. false

15. A normally ____________ switch sits on a shelf with its contacts connected.

16. A tachometer is a sensor.

a. true
b. false

17. A variable-speed drive is a type of __________ used in newer HVAC systems to boost efficiency.

18. A series of steps taken by hardware is called a(n)

a. program.
b. algorithm.
c. instruction set.
d. command line interface.

19. ____________ is the time it takes to physically move a switch when given a command.

a. Switching time
b. Propagation delay
c. Control lag
d. Reaction time

20. Discrete systems rely on ____________ components, while integrated systems rely on ____________ ones.

a. combined, individual
b. individual, combined
c. separate, high-power
d. low-power, high-power

21. ____________ energy, or energy in motion, and is related electrically to ____________.

a. Kinetic, voltage
b. Potential, current
c. Kinetic, current
d. Potential, voltage

22. PLCs and security systems communicate using voltages between ________ and ________.

23. In typical electrical circuits, natural frequency is 60Hz.

a. true
b. false

24. A(n) ____________ is a device that uses low power signals to control large electrical loads.

25. A(n) ______________ transformer increases available current.

a. step-up
b. current gain
c. unity equalizer
d. step-down

26. Which type of switch senses the presence or absence of a part touching its surface?

a. limit
b. hardware
c. contact
d. tongue

27. The current status of a system is called ______________, or scene.

28. Digital systems can only ask and interpret yes or no questions and responses.

a. true
b. false

29. A(n) ______________ system or parameter does not change, while a(n) ______________ system or parameter does.

30. The NEC maintains no jurisdiction over most industrial installations.

a. true
b. false

Posttest 1

YOU ARE NOW FAMILIAR with the kinds of questions and answer formats you will see on an electrician's licensing exam. Take this posttest to identify any areas that you may need to review in more depth before the test day. When you are finished, check the answers on page 262 carefully to assess your results. Each answer in the Answers and Explanations section provides a cross reference to the electrical code addressed in the question. Use the correlation guide in Appendix A to connect these codes to specific chapters in this book as well as in the NEC codebook. Remember to do the following:

- Work carefully.
- Use estimation to eliminate answer choices or to check your work.
- Answer every question.
- Check to make sure your answers are logical.

To simulate the test conditions, use the time constraints of an official electrician's licensing exam. Remember, on the official exam, an unanswered question is counted as incorrect, so it's always better to make a good guess.

1. According to NEC, 18AWG copper conductor protection cannot exceed ________ amperes.
a. 2
b. 5
c. 7
d. 10

2. Equipotential bonding is installed primarily to ________ voltage gradients in a pool area.

3. A hybrid electrical system includes such energy storage systems as batteries, flywheels, and superconducting magnetic storage equipment.
a. true
b. false

4. The acronyms SCADA, COPS, and DCOA are defined in NEC article ________.

5. An aircraft hangar, including any adjacent, connecting areas, is classified as a Class 1, Division 2 or Zone 2 location up to ________ inches above the floor.
a. 12
b. 18
c. 24
d. 48

6. A dining room circuit must use ________ wire.

7. Underground wiring in a warehouse must be routed in metal conduit when less than 24 in. deep.
a. true
b. false

8. A bathroom with a window requires an exhaust fan.
a. true
b. false

9. NEC outlines requirements for which of the following?
a. underground mining installations
b. ships and nautical wiring
c. communication facilities
d. fiber optic cabling

10. An authority having jurisdiction (AHJ) can waive portions of NEC.
a. true
b. false

11. "The overall covering shall be flame retardant, moisture resistant, fungus resistant, and corrosion resistant" best describes the cover on ________ cable.

12. When the demand factors of NEC Table 555.12 apply, the demand factor in NEC 220.61(B) is also allowed.
a. true
b. false

13. According to NEC ________, all panel boards require circuit directories.

14. Which of the following cannot source a portable left foot accelerator (PLFA) power supply?
a. three-phase circuit
b. AFCI protected circuit
c. GFCI protected receptacle
d. gas generator

15. Electricians should consider all electrical equipment and supplies inactive until they determine otherwise.

a. true

b. false

16. NEC article ________ covers metallic raceways, while NEC article ________ covers and nonmetallic raceways.

17. The maximum length for a kitchen disposer cord is

a. 36 in.

b. 24 in.

c. 12 in.

d. 6 in.

18. A(n) ________ is a device that interfaces power to a load.

19. In NEC, which table helps determine the minimum allowable ampacity of THHR?

a. 310.16

b. 402.3

c. 620.14

d. 705.3

20. Overhead service may use PVC pipe.

a. true

b. false

21. Central heating equipment is considered an appliance.

a. true

b. false

22. For a portable motor rated at ________ horsepower or lower, the controller can be an attachment plug and receptacle.

a. $\frac{1}{4}$

b. $\frac{1}{8}$

c. $\frac{1}{3}$

d. $\frac{1}{2}$

23. By code, ________ receptacles can reside on a 20-amp circuit.

24. Unless identified for such locations, explosion-proof equipment and wiring are unacceptable in Class 2 locations.

a. true

b. false

25. The ampacity of capacitor circuit conductors should not be less than ________% of the capacitor's rated current.

a. 80

b. 90

c. 120

d. 135

26. The symbol [symbol] represents which electrical device?

____

____

____

____

27. Snap switches for inductive loads must be rated at ISA or greater.

a. true

b. false

28. A high-impedance grounded system can only be installed when continuous service is needed.
a. true
b. false

29. Branch-circuit overcurrent device ratings should not exceed appliances' protective device ratings.
a. true
b. false

30. When in parallel, conductors must match in all ways EXCEPT
a. AWG.
b. insulation color.
c. composition.
d. terminations.

31. In air-conditioning and refrigerating equipment, the branch-circuit selection current equals or exceeds the marked rated-load current.
a. true
b. false

32. In bus bars and conductors for motor circuits and controllers, the ________ phase has the higher voltage to ground on 3-phase, 4-wire, delta-connected systems.

33. Chain-supported luminaires in show windows can be externally wired.
a. true
b. false

34. Electricians who need more than 6' of liquid-tight flexible nonmetallic conduit should reference NEC section ________.

35. In a recreational vehicle, a garbage disposal requires a cord.
a. true
b. false

36. According to NEC, "All circuit conductors between the service equipment, the source of a separately derived system, or other power supply source and the final branch-circuit overcurrent device" is known as the ________.

37. The ____________ type of pipeline heater uses hysteresis.
a. resistance
b. skin effect
c. induction
d. impedance

38. NEC 314.27(A) and (B) contain the most stringent requirements for ceiling fans.
a. true
b. false

39. According to NEC, ________ receptacles are needed outside of a residence.

40. The best way to ground computer and data-processing equipment is to bond the receptacle ground to the metal raceways directly.
a. true
b. false

41. NEC section ________ covers requirements for permanently connected appliances.

42. The maximum operating temperature for Thermoset is

a. 32°F.
b. 60°F.
c. 75°F.
d. 90°F.

43. Cables may not be installed using directional boring.

a. true
b. false

44. Which of the following articles permits non-metallic cable branch circuits to be installed temporarily in structures lacking height limitations?

a. 334.10(3)
b. 590.4(C)
c. 830.40(B)
d. none of the above

45. In essence, the NEC is a "best practices" handbook for quality electrical work.

a. true
b. false

46. The fill requirements for ENT appear in NEC ________.

47. The definition for the electrical phrase *in sight* appears in NEC Section

a. 220.51.
b. 240.6.
c. 312.11.
d. 430.102.

48. In electrical equipment rooms, illumination must always be automatically controlled.

a. true
b. false

49. Citing NEC references, describe the process for determining the number of conductors that can be carried in an existing raceway.

____

____

____

____

50. In electric circuits, a load usually converts electrical energy to ____________, ____________, or ____________.

51. A structure intended to mount a wet-niche luminaire in a pool or spa is called a

a. wet mount.
b. forming shell.
c. moisture barrier.
d. moisture shunt.

52. A sump pump requires a dedicated circuit.

a. true
b. false

53. In electrical terms, all of the following qualify as patient-care areas EXCEPT

a. examining rooms.
b. nursing home sleeping rooms.
c. wet procedure spaces.
d. treatment rooms.

54. Explosion properties are the same as ignition properties.
a. true
b. false

55. A receptacle or switch receiving 14-2 wire counts as ________ cubic inches.
a. 2
b. 4
c. 7
d. 14

56. PVC expands more per degree increase in temperature than RTRC.
a. true
b. false

57. The bare wire arising from the power-company side of electrical service is considered ____________.

58. Electricians should apply knob-and-tube wiring only when working
a. on new installations.
b. in commonwealths.
c. on expansions or repairs.
d. all of the above

59. Electricians can install feeders with a common neutral.
a. true
b. false

60. Splices in knob-and-tube wiring should be ____________.

61. Window candle-switch outlets in side bathrooms require GFI protection.
a. true
b. false

62. In NEC, Section ________ covers how to ground equipment that is wired permanently, while NEC Section ________ outlines cord and plug wiring methods.

63. When more than one receptacle or load is on the branch circuit for a recreational vehicle, a 15-ampere receptacle can be protected by a 20-ampere fuse or circuit breaker.
a. true
b. false

64. Since fluorescent lamps are "cool" as compared to incandescent lamps, the NEC does not require thermal protection for fluorescents.
a. true
b. false

65. NEC requires electricians to staple conduit at 5 ft. intervals.
a. true
b. false

66. Unused switchboard openings require no protection.
a. true
b. false

67. In NEC, Table ________ lists the minimum sizes of equipment grounding conductors for grounding raceway and equipment.

a. 240.33
b. 250.122
c. 326.80
d. 620.14

68. NEC 2008 disallows the use of iron as a grounding electrode.

a. true
b. false

69. NEC suggests that a system's design specification accommodate expansion.

a. true
b. false

70. Four parties who should be satisfied with the quality of electrical work in a new residence are the ____________, ____________, ____________, and ____________.

71. Range hoods must be hardwired.

a. true
b. false

72. The top of a switchboard must be at least 18 in. below the ceiling.

a. true
b. false

73. NEC Article 450 covers the installation of all transformers.

a. true
b. false

74. Hydrogen is a Class ________ Group ________ material.

a. 1, A
b. 2, A
c. 1, B
d. 2, B

75. Service lateral conductors may not be tapped or spliced.

a. true
b. false

76. A feeder supplying 15- and 20-ampere receptacle branch circuits can be protected by a ground-fault circuit interrupter in lieu of the provisions for such interrupters as specified in NEC

a. 210.8 and 590.(6)(A).
b. 210.52.
c. 210.8.
d. 210.8, 210.52 & 590.(6)(A).

77. In Table ________, Sections ________, NEC addresses grounding in floating buildings.

78. Any carnival equipment that the general public can access must be GFCI protected.

a. true
b. false

79. When calculating continuous and noncontinuous loads in amperes, electricians should consult which NEC article?

a. 550
b. 314
c. 430
d. 109

80. When seven elevators are installed on one feeder, the demand factor should be

a. 1.00.
b. 0.92.
c. 0.88.
d. 0.77.

81. A meter can measure current just outside a circuit.

a. true
b. false

82. A GFCI will trip in ______________.

83. A load that receives power for at least three hours at a time is considered

a. contiguous.
b. half-day.
c. continuous.
d. triple.

84. Smoke detectors can reside on separate circuits.

a. true
b. false

85. When an attic has a furnace, that attic requires a receptacle.

a. true
b. false

86. According to NEC 310.11, conductors must be marked with such information as ______________, ______________, ______________, and ______________.

87. A ground ring is installed _________ below the earth's surface.

a. 12 in.
b. 18 in.
c. 30 in.
d. 36 in.

88. NEC identifies the required height of dryer plug receptacles.

a. true
b. false

89. The symbol [symbol] represents a(n) ______________, which is usually located ______________.

90. Floor receptacles are acceptable for residential use.

a. true
b. false

91. According to the principle of ____________ ____________, a disconnect is the main panel while the panel in the residence is called a subpanel.

92. The S_3 notation indicates which of the following?

a. three-way switch
b. triangle receptacle
c. sole side switch
d. none of the above

93. Wire that is _________ AWG must feed a 200-amp breaker.

94. The circuit for a GFCI bathroom receptacle is how many amps?

a. 60
b. 55
c. 30
d. 20

95. Overcurrent devices protect all types of devices.

a. true
b. false

96. Aluminum conductors cannot terminate within ________ from the earth.

a. 12 in.
b. 18 in.
c. 36 in.
d. 48 in.

97. The NEC section referenced most often when calculating loads for farm applications is ________.

98. An 18' wall with a door at either end requires ________ receptacles.

99. Which of the following NEC sections covers ungrounded systems?

a. 110.14(C)
b. 250.4(B)
c. 501.15(A)
d. 668.21(D)

100. With advances in electrical coding, junction boxes no longer reside behind drywall.

a. true
b. false

101. The requirements for electrical connectors appear in NEC Section ________.

102. When a transformer has a voltage ratio of 8:1 and an input of 120V, the output voltage is

a. 5.
b. 15.
c. 60.
d. 120.

103. All splices in electrical wiring must be covered with

a. duct tape.
b. a nonmetallic plate.
c. cloth electrical tape.
d. insulation with the same rating as that on the conductor being spliced.

104. The entity known as ____________ ____________ determines the maximum ampacity of a circuit.

105. A surface raceway may extend through a partition when the raceway is a single piece.

a. true
b. false

106. How many wires are needed from a main panel to a subpanel in the same building?

a. two
b. three
c. four
d. six

107. The 2' of countertop behind a sink requires a receptacle.

a. true

b. false

108. An exhaust fan and light in a shower area require GFI protection.

a. true

b. false

109. Residential dwellings may use type ________, ________, and ________ cables.

110. According to NEC Section(s) ________, electricians may not exceed 75% of box fill when splicing and taping conductors in a box.

a. 386.56

b. 388.56

c. 398.3

d. a and b

111. The spacing requirements for receptacles on wall space is found in NEC ________.

112. There are only seven permissible ways to connect grounding and bonding equipment.

a. true

b. false

113. For determining farm load demand factor, use NEC Table

a. 210.2.

b. 220.103.

c. C.11(A).

d. none of the above

114. In a Class ________, Division ________, location, dust may, in unusual circumstances, present an explosion hazard.

a. I, I

b. II, I

c. II, II

d. III

115. According to NEC 250.30, when a portable optional standby source is used as a separately derived system, it shall be grounded to a grounding electrode.

a. true

b. false

116. Eight locations requiring GFCI circuit protection appear in NEC Section ________.

117. Branch circuits must be loaded equally in terms of wattage per unit area.

a. true

b. false

118. Electricians who are running isolated receptacles for computers should consider NEC Section

a. 250.146(D).

b. 312.3.

c. 430.113.

d. 550.4(A).

119. All balconies and decks that are accessible from inside dwellings require receptacles.

a. true

b. false

120. All outside fixtures that get wet or damp should be marked for wet locations.

a. true
b. false

121. Which of the following NEC articles establishes the requirements for fixed electrical space heating equipment?

a. 710
b. 556
c. 424
d. 201

122. Underground meter bases must be ________ high from finish grade.

123. NEC Section 408.4 requires that unused panelboard overcurrent devices be labeled as such.

a. true
b. false

124. NEC sections ________, ________, and ________ cover sufficient cabinet clearances for bending and splicing of conductors.

125. Split receptacles can be installed in locations like ____________, ____________, and ____________.

126. According to NEC Section ________, a light is required in all working spaces with electrical equipment.

a. 510.26
b. 434.4.2
c. 110.26.2
d. 107.4

127. NEC Article 220 identifies minimum lighting loads for given types of occupancy.

a. true
b. false

128. Fiber optic cabling appears in NEC Article

a. 380.
b. 600.
c. 770.
d. 821.

129. In a residence, areas like the ____________, ____________, and ____________ require 20-amp circuits.

130. An enclosure supported by a pendant need not comply with NEC 314.23(H)(1) or (H)(2).

a. true
b. false

131. Electricians who are calculating loads for receptacles on the trap should refer to which NEC section?

a. 220.3
b. 312.10
c. 430.103
d. 520.4

132. Frames of switchgear and control assemblies shall be connected to a(n) ____________ or, where permitted, the grounded conductor.

a. current-limiting protective device
b. nonmetallic base
c. insulating material
d. equipment grounding conductor

133. Crawl spaces require lights when used for storage.
a. true
b. false

134. An electric dryer needs a neutral to ____________.

135. Which NEC articles and tables address transformer installations into two voltage levels?

____

____

____

____

136. Wires extending from a residence to a post light not protected by a GFI must be buried how deep?
a. 18 in.
b. 24 in.
c. 3 ft.
d. 4 ft.

137. Liquid-filled transformers are subject to the same requirements regardless of the liquid they use.
a. true
b. false

138. NEC Article 526 helps electricians determine requirements for
a. embedded deicing systems.
b. open or exposed deicing systems.
c. a and b
d. none of the above

139. Variable-speed motors require protection for ____________ and overcurrent.

140. According to NEC 230.71(A), as many as six disconnects may be used to turn off power to a building.
a. true
b. false

141. On a circuit using a 20-amp breaker and 12 AWG wire, a 15-amp receptacle is acceptable.
a. true
b. false

142. When installing a 200 watt bulb in a 300 watt lighting fixture, ________ watts are used.

143. In compliance with NEC Section ________, motor overload protection is based on the motor nameplate rating.
a. 313.44(B)
b. 430.32
c. 517.23(C)
d. all of the above

144. A surge arrester over 1 kV can be installed when the rating of the surge arrester is less than the maximum continuous phase-to-ground power frequency voltage available at the point of application.
a. true
b. false

145. Provisions for sensitive electronic equipment as discussed in NEC Article ________ apply only to industrial installations.

146. Receptacles as specified in NEC Article 647 may not be used for

a. motors.
b. lighting.
c. telecom equipment.
d. computers.

147. Panelboards and switchboards differ in that switchboards are designed for higher ____________.

148. NEC Article ________, Part ________ governs cartridge fuses.

a. 280, 4
b. 315, 2
c. 240, 3
d. 240, 6

149. Electricians may opt to use no grounding electrode when a suitable substitute is available.

a. true
b. false

150. Smoke detectors must reside on AFI breakers.

a. true
b. false

151. The requirements for audio-processing equipment are outlined in which NEC chapter?

a. 5
b. 6
c. 7
d. none of the above

152. NEC Section ________ provides requirements for track lighting.

153. NEC Article ________ governs the installation of remote controls and signal circuits.

a. 330
b. 525
c. 725
d. 730

154. Class I power source limitations for alternating current appear in NEC Chapter ________, Tables ________ and ________.

155. Cable running under a building within a raceway must be buried 36 inches.

a. true
b. false

156. When installing low-voltage wires with a 120 volt power line, electricians must go to NEC Section ________ to meet the compliance.

a. 725.54
b. 525.54
c. 327.23
d. 527.23

157. This symbol [symbol] represents what concept?

____

____

____

____

158. An electric water heater is considered an intermittent or noncontinuous load when calculating branch circuit ampacities.

a. true
b. false

159. NEC Article ________ covers requirements for air-conditioning and refrigeration equipment.

160. Installing a branch circuit over current devices with a rating not to exceed the MOP is covered in NEC Article ________, Section ________.

a. 340, 440.4
b. 440, 240.6
c. 540, 340.4
d. 440, 250.6

161. NUCC can be used inside buildings.

a. true
b. false

162. NEC Section ________ states that the Earth is not to be the only equipment grounding conductor.

163. In NEC terminology, the phrase *shall be* means ____________.

164. Which of the following insulated conductor types is allowed in dry/damp locations?

a. PFA
b. THHW
c. FEP
d. all of the above

165. NEC Tables 310.67 through 310.86 do NOT provide information on ampacities and conductor

a. length.
b. gauge.
c. diameter.
d. temperature.

166. NEC Section ________ covers equipment grounding conductors for copper aluminum or copper-clad aluminum and several types of conduit and metallic tubing.

167. The major cause of death from electrical shock is heart fibrillation.

a. true
b. false

168. The smallest counter that needs a receptacle is

a. 8 in.
b. 12 in.
c. 14 in.
d. 18 in.

169. The types of insulation material deemed acceptable in wet locations appear in NEC Article ________.

170. Unless otherwise provided, Type AC cable should be secured within ________ in. of every outlet box, junction box, cabinet, or fitting and at intervals not exceeding ________ ft. where installed on or across framing members.

a. 8, $4\frac{1}{2}$
b. 12, $4\frac{1}{2}$
c. 18, 4
d. 16, 4

171. Electricians must run 14-3 between smoke alarms to keep the devices in proximity.

a. true
b. false

172. Which of the following kitchen receptacles need(s) no GFCI protection?

a. stove
b. garbage disposal
c. refrigerator
d. all of the above

173. Bedroom circuits may use 12-2 Romex.

a. true
b. false

174. Branch circuit ampacity must not be less than ________% of the continuous load plus ________% of the noncontinuous load.

175. Devices run at 240V as opposed to 120V offer the benefit of

a. high voltage.
b. low amperage.
c. low power loss.
d. all of the above

176. Leads to an unused current transformer should never be short circuited.

a. true
b. false

177. Only someone on the code-making panel (CMP) may submit proposals to amend NEC.

a. true
b. false

178. NEC Chapter ________, Table ________ covers the resistance of copper and aluminum wire.

179. A(n) ____________ conductor applies as a voltage ground, not a grounding path.

180. Valance lighting fixtures are located where?

a. floor level
b. eye height
c. close to a ceiling
d. in a corner

181. The maximum motor horsepower that can be disconnected using the panel breaker is

a. 0.1.
b. 0.125.
c. 0.25.
d. 1.0.

182. AFCI is used to protect ____________, while GFCI is used to protect ____________.

183. Overhead conductors that are outdoors must be protected with coverings and/or insulated for a distance of ________ beyond the building.

a. 5 ft.
b. 6 ft.
c. 10 ft.
d. 25 ft.

184. Continuous loads are defined in NEC Article ________.

185. The Earth is an efficient ground fault current path.
- **a.** true
- **b.** false

186. Open conductors on separate circuits must be separated by
- **a.** 1 in.
- **b.** 50 mm.
- **c.** 4 in.
- **d.** 200 mm.

187. Ground fault current path is covered under NEC Section ________.

188. In a residence, each square foot should be allocated ________ VA.
- **a.** 1
- **b.** 2
- **c.** 3
- **d.** 5

189. Electrically conductive materials that are likely to become energized must be connected.
- **a.** true
- **b.** false

190. Most local codes require ________ ground rods for service that are ________ long and ________ apart.

191. Overhead service clearance requirements can be found in NEC Article
- **a.** 250.
- **b.** 230.24.
- **c.** 720.
- **d.** 547.

192. Electricians can use cable ties to support new LV wiring in existing conduit.
- **a.** true
- **b.** false

193. According to NEC Section ________, circuits need not be grounded.

194. Fusing the load side of a transformer suffices to protect the transformer itself.
- **a.** true
- **b.** false

195. Handhole enclosures used in underground systems should be
- **a.** identified.
- **b.** buried at least 36 inches.
- **c.** provided with a manufactured bottom plate.
- **d.** sized for personnel to enter.

196. Temporary service must have a manual disconnect rated and installed under the requirements that apply to permanent service.
- **a.** true
- **b.** false

197. According to NEC Section ________, an enclosed service disconnect must be bonded in the enclosure.

198. At 1,000 or less, voltage must run to each service disconnecting means and bond to each disconnecting enclosure.
- **a.** true
- **b.** false

199. Clearances for overhead conductors apply to ____________ level.

200. NEC Article ________, Part ________ provides guidance for installing a cable-type permanent space heater.
- **a.** 524, 5
- **b.** 514, 5
- **c.** 424, 5
- **d.** 324, 4

201. The structure of a steel building may be used to augment a grounding system with too much resistance.
- **a.** true
- **b.** false

202. Which of the following NEC articles covers circuit protective devices?
- **a.** 100
- **b.** 200
- **c.** 240
- **d.** 500

203. NEC Article 285 fails to recognize TVSS as a protective device at less than 1,000 V.
- **a.** true
- **b.** false

204. With some exceptions, only conductors of size ________ may be installed in parallel.

205. Which of the following NEC tables would an electrician use to choose the grounding electrode conductors for a service where there is no outer current protection ahead of the service other than the utility company?
- **a.** 310.8
- **b.** 705.3
- **c.** 680.8
- **d.** 250.65

206. A 4' kitchen island requires a receptacle.
- **a.** true
- **b.** false

207. The wire that goes on the common screws of a three-way switch is
- **a.** black.
- **b.** white.
- **c.** green.
- **d.** clear.

208. Because it resides partially in the ground, a metal fence requires no connection to a ground rod.
- **a.** true
- **b.** false

209. When electricians cannot limit fill rate to 20% using a metallic wireway, a nonmetallic wireway can be filled to more than 20% capacity.
- **a.** true
- **b.** false

210. It is proper protocol to install an OCPD in series with each

a. circuit breaker.
b. ungrounded conductor.
c. three-phase leg.
d. grounded conductor.

211. At most, a service drop or lateral may supply ________ service conductors.

212. Switches must be installed with the "off" position up and the "on" position down.

a. true
b. false

213. When running tap conductors from a box to a light fixture, the conditions in NEC Section ________ must be met.

214. Constant wattage heating cables should not exceed ________ watts/linear foot of cable when installed in concrete or masonry floors.

a. $16\frac{1}{2}$
b. $15\frac{1}{2}$
c. $12\frac{1}{2}$
d. $10\frac{1}{2}$

215. In no case shall the volume of a box, as calculated per NEC 314.16(A), be less than the fill, as calculated per NEC 314.16(B).

a. true
b. false

216. The wire on kitchen counter circuits must be ________ AWG.

217. A refrigerator circuit counts as a counter circuit.

a. true
b. false

218. NEC Section ________ outlines the requirements for installing parking lot lights around a building.

219. A separate insulated isolated equipment grounding conductor is sized according to NEC Table

a. 225.61.
b. 250.122.
c. 300.19(A).
d. 310.13(A).

220. Electricians can place lighting circuits in more than one room.

a. true
b. false

221. A kitchen island requires a receptacle when

a. any side of the island is longer than 3 ft.
b. the island's surface area is 2 sq. ft. or more.
c. the island has a sink.
d. none of the above

222. The computed load of a feeder must exceed the computed load of the circuits it feeds.

a. true
b. false

223. A device box displays its cubic inches on its ____________ surface.

224. Electricians should install services to use ________ of a breaker.

a. 30%
b. 50%
c. 80%
d. 100%

225. Five locations in a residence that require GFCI protection are the ________________, ________________, ________________, ________________, and ________________.

226. NEC Article ________ provides a backdrop for the scope and application of NEC itself.

a. 310
b. 270
c. 220
d. 90

227. Vending machines require GFCI protection.

a. true
b. false

228. NEC Article ________ provides guidelines for determining motor ampacity requirements.

229. Multiwire branch circuits are acceptable in Class 1 environments.

a. true
b. false

230. A PLC may serve as the primary disconnecting means for a pipe heating sleeve.

a. true
b. false

231. The person responsible for starting, stopping, and controlling an amusement ride or supervising a concession is the

a. operator.
b. jurisdiction.
c. guard.
d. controller.

232. A 1 HP motor at 208V has a full load current draw of

a. 16 amps.
b. 9.2 amps.
c. 8.8 amps.
d. 8.0 amps.

233. Requirements for transformer ventilation appear in which of the following NEC sections?

a. 225.3
b. 240.4
c. 300.1
d. 450.9

234. NEC 314.39 addresses the guarding of live or energized components.

a. true
b. false

235. The "three-foot rule" for clearances around devices like control panels is a

a. minimum that must be equaled or exceeded.
b. maximum that must not be exceeded.
c. a and b
d. none of the above

236. In a three-phase delta system, the leg with the highest voltage must be

a. reduced to within 1% of other leg voltages.
b. considered the B leg.
c. permanently labeled as HV.
d. connected directly to neutral.

237. A floor drain in a vault intended for wet transformer(s) is good practice regardless of rules or requirements.

a. true
b. false

238. The handles or escutcheon areas of circuit breakers rated at ________ amperes or less and ________ volts or less should be molded, stamped, etched, or similarly marked with their ampere ratings.

a. 75, 300
b. 100, 600
c. 80, 300
d. 90, 600

239. A TVSS provides adequate protection against lightning strikes.

a. true
b. false

240. NEC 210.52 states that "Receptacles shall be installed so that no point measured horizontally along the floor line in any wall space is more than 1.8 meters (6 ft.) from a receptacle outlet," which means that, along a long wall, receptacles must be within ________ ft. of each other.

241. In a commercial environment where a branch circuit could effectively handle 20 KW, demand would be calculated as ________ W.

a. 15,000
b. 12,000
c. 10,000
d. 8,000

242. Which of the following entities does NOT require AFCI protection?

a. branch circuit where the initial run is encased in AC up to the first, plastic-mounted receptacle
b. clothes closet measuring more than 10 square feet
c. fire alarm system installed in EMT using metal boxes
d. sunroom in a typical residence

243. NEC Table 250.66 applies to the derived conductors of separately derived AC systems.

a. true
b. false

244. Which one of the following statements about spas and hot tubs is NOT accurate?

a. Receptacles should at least 3 ft., measured horizontally, from the inside walls of the spa or hot tub.
b. Receptacles that are rated 125 volts and 30 amperes or less and that are within 10 ft. of the inside walls of a spa or hot tub require GFCIs.
c. Switches should be at least 5 ft., measured horizontally, from the inside walls of the spa or hot tub.
d. all of the above

245. Circuit breakers can be run by electric or pneumatic methods if manual means are available.

a. true
b. false

246. Fixed electric space-heating loads should be calculated at ________% of the total connected load.

247. When used as listed, the grounding terminal of a hospital-grade receptacle must connect to ground by which of the following?

a. bare wire
b. wire one gauge larger than that used for the other conductors
c. insulated wire
d. any of the above

248. A(n) ____________________ is someone with skills and knowledge related to the construction and operation of equipment, as well as training in the related hazards.

a. electrical inspector
b. AHJ representative
c. job foreperson
d. qualified person

249. The motors, controllers, and wiring of electrically operated pool covers should be no less than ________ from the inside wall of the pool unless they are separated from the pool by a wall, a cover, or another permanent barrier.

a. 6 ft.
b. 5 ft.
c. 2 ft.
d. 1 ft.

250. A procedure table in a critical care area is considered a patient bed location.

a. true
b. false

Posttest 2

YOU ARE NOW FAMILIAR with the kinds of questions and answer formats you will see on an electrician's licensing exam. Take this posttest to identify any areas that you may need to review in more depth before the test day. When you are finished, check the answers on page 272 carefully to assess your results. Each answer in the Answers and Explanations section provides a cross reference to the electrical code addressed in the question. Use the correlation guide in Appendix A to connect these codes to specific chapters in this book as well as in the NEC codebook. Remember to do the following:

- Work carefully.
- Use estimation to eliminate answer choices or to check your work.
- Answer every question.
- Check to make sure your answers are logical.

To simulate the test conditions, use the time constraints of an official electrician's licensing exam. Remember, on the official exam, an unanswered question is counted as incorrect, so it's always better to make a good guess.

1. Which of the following is NOT considered a fountain?
   a. reflecting pool
   b. display pool
   c. drinking fountain
   d. all of the above

2. In livestock housing, the part of the direct-buried equipment grounding conductor that emerges from the ground on its way to the building must be insulated or covered copper.
   a. true
   b. false

3. NEC Article ________ covers grounding electrode systems.

4. Switches must be marked with three ratings: ____________, ____________, and ____________.

5. NFPA 70 differs from NEC in that NFPA 70 addresses fire safety while NEC addresses electrical safety.
   a. true
   b. false

6. When considering an effective grounding path, which of the following statements is true?
   a. Insulated joints should be bonded.
   b. Any equipment likely to be disconnected for repairs or replacement should be bonded.
   c. Bonding jumpers should be long enough to allow equipment removal while retaining the grounding path's integrity.
   d. all of the above

7. Property owners can act as their own authorities having jurisdiction (AHJs).
   a. true
   b. false

8. According to NEC Section ________, electricians may replace defective, ungrounded receptacles with like replacements.

9. NEC prevents the installation of inductive electric heaters.
   a. true
   b. false

10. The branch circuit conductors supplying one or more units of a data-processing system should have a(n) ____________ not less than ________% of the total connected load.
    a. ampacity, 125
    b. voltage drop, 3.5
    c. harmonic distortion, 5
    d. operating ratio, 1.5–2.5

11. All metallic boxes and components must be grounded, unless a grounding conductor is absent.
    a. true
    b. false

12. A recessed luminaire not identified for contact with insulation shall be not less than ________ from combustible materials.
    a. $\frac{1}{4}$ in.
    b. $\frac{1}{2}$ in.
    c. $\frac{3}{4}$ in.
    d. $\frac{5}{8}$ in.

13. Intermediate metal conduit (IMC) with a 1 in. trade size has a metric designator of

a. 27.
b. 28.
c. 30.
d. 32.

14. All metal boxes require bonding with ground screws and clips.

a. true
b. false

15. Secondary electrical distribution systems to mobile-home lots should be

a. single-phase 120/240, nominal.
b. three-phase 120/208, nominal.
c. a and b
d. none of the above

16. Finished basements require GFCI protection.

a. true
b. false

17. Class ________ locations have hazardous combustible dust.

18. In a parallel circuit, each load is connected in a separate path.

a. true
b. false

19. Trade size $\frac{1}{2}$ ENT has a nominal internal diameter of

a. 0.5125 in.
b. 0.5250 in.
c. 0.5625 in.
d. 0.560 in.

20. A raceway and its terminations must be in areas at the same temperature.

a. true
b. false

21. A 10 AWG copper wire holds ________ amps.

a. 10
b. 20
c. 30
d. 50

22. Expansion contraction is not an issue when heating elements and assemblies bridge expansion joints.

a. true
b. false

23. The symbol represents the device known as the ________________.

24. NEC Article ________ outlines the requirements for construction and panel application.

25. NEC discusses how to size a DC bonding jumper yet does not require that one be installed.

a. true
b. false

26. When adding lighting, the electrician's first step should be to

a. calculate the number and locations of receptacles.
b. determine the added load the supplying circuit can deliver.
c. run the circuit and service panel at full capacity.
d. none of the above

27. NEC Sections ________ and ________ address rooftop receptacle outlets.

28. Electricians can install breakers in panel boxes that differ from the manufacturers' panels by code.
   a. true
   b. false

29. SPD devices must be supplied with ________ AWG copper or aluminum.

30. Thermal protection is required in a recessed luminaire installed in poured concrete.
   a. true
   b. false

31. A storage battery in a legally required standby system must source ________% of nominal voltage.
   a. 50.0
   b. 75.0
   c. 82.5
   d. 87.5

32. In a(n) ____________ type of circuit, all current flows through all loads.

33. Secondary circuits for lighting systems operating at 30 volts or less should be grounded.
   a. true
   b. false

34. NEC Article ________ covers air conditioners and equipment with hermetic refrigeration.
   a. 440
   b. 450
   c. 610
   d. 625

35. To ensure safety when wiring a hot panel, electricians should ________________________, ________________________, and ________________________.

36. Which of the following NEC tables summarizes branch circuit requirements?
   a. 830.47
   b. C.11.7
   c. 210.24
   d. B.310.9

37. A kitchen requires ____________ circuits.

38. NEC Article 310 provides the temperature ratings of wires' outer jackets.
   a. true
   b. false

39. A continuous-duty power resistor conductor must be rated at ________% of full-load current.
   a. 60
   b. 80
   c. 110
   d. 125

40. As a piece of wire's diameter increases, its AWG value increases.
   a. true
   b. false

41. ____________ type conductors are not rated for a 90 C (194F) maximum operating temperature.
   a. THW-2
   b. ZW-2
   c. USE
   d. RHH

42. NEC Article ________, Part ________ contains the requirements for motor protection.

43. According to NEC, EMT use is allowed for which of the following?
a. corrosion protection using enamel
b. exposed work
c. luminaire support
d. contact with dissimilar metals

44. NEC Table 310.16 is commonly used to temperature rate wires.
a. true
b. false

45. Hallways that are 10-ft. long require receptacles every ________ ft.

46. Per NEC Section ________, electricians who are calculating transformer feeder ampacity should use transformers' nameplate ratings.
a. 660.4
b. 554.1
c. 334.2
d. 215.2

47. The motor on a hot tub requires GFCI protection.
a. true
b. false

48. The interior of a raceway that installed in a wet location is itself a wet location.
a. true
b. false

49. When installing conductors in a raceway or cable, electricians should consult NEC Section
a. 310.15(B)(2).
b. 344.10(A)(1).
c. 410.130(F)(2).
d. none of the above

50. Per NEC 360.12, ____________ type raceways are disallowed from battery storage areas.

51. ENT must be marked at intervals not to exceed
a. 12 ft.
b. 10 ft.
c. 6 ft.
d. 3 ft.

52. A 100-ft. piece of RTRC that is cut at 20° will be ________ ft. ________ in. at 90°.

53. Intermediate IMC may not be used as a grounding conductor.
a. true
b. false

54. NEC Articles 517 and 660 are similar in that both address
a. residential kitchens.
b. industrial heat protection.
c. x-ray machines.
d. none of the above

55. Lightning protection system ground terminals should be bonded to the building or structure grounding electrode system.
a. true
b. false

56. Special requirements for lighting in closets exist because
a. clothes closets are considered moist areas.
b. clothes closets require diffused lighting.
c. clothes closets are small and concentrate heat.
d. There are no such requirements.

57. According to NEC Section ________, a number 6 AWG or larger ground must carry a white or gray termination point.

58. A fixed space heater requires a disconnect.
a. true
b. false

59. ENT must be supported every ________ ft. and at ________ ft. from a box.
a. 3, 1.5
b. 3, 3
c. 6, 3
d. 6, 1.5

60. All new wiring receptacles must be grounded to a neutral bar in the panel box.
a. true
b. false

61. Single-throw knife switches must be installed so that ____________ cannot ____________ them.
a. children, reach
b. gravity, open
c. gravity, close
d. inspectors, find

62. Direct-current motors are always protected at a maximum of ________% of the full load current.

63. The smallest residential service is ________ amps.

64. Short-circuit and ground-fault protective devices for motors should be based on circuit amperage.
a. true
b. false

65. According to NEC, 120V track lighting should be mounted at least ________ above the floor.
a. 4 ft.
b. 5 ft.
c. 8 ft.
d. 10 ft.

66. Audible and visual signal devices, except portable standby power sources, should be provided, where practicable, for derangement and load carrying.
a. true
b. false

67. Per NEC Section ________, a hidden device terminal intended for a grounding conductor must be marked green, "GR," or "G," or carry a grounding symbol.

68. In a refrigeration system that has a first-floor walk-in room and basement compressors, the motors in the two assemblies are considered one unit.
a. true
b. false

69. NEC Article ________ covers IMC and describes its permitted uses.

a. 610
b. 560
c. 432
d. 342

70. NEC provides no requirements for device voltage drops.

a. true
b. false

71. A 9 ft. countertop requires how many receptacles?

a. one
b. two
c. three
d. eight

72. In a poured concrete floor, heating elements are spaced at ________ minimum on-center.

a. $\frac{1}{2}$ in.
b. 1 in.
c. 2 in.
d. 2.5 in.

73. Per NEC 285.4, one SPD can be used to protect ungrounded conductors when rated at ________ V or less.

74. GE panels can accept Homeline breakers.

a. true
b. false

75. Per NEC Article ____, for ungrounded circuits, voltage to ground is defined as the greatest voltage between the given conductor and any other conductor of the circuit.

a. 100
b. 110
c. 200
d. 310

76. Generator locations must meet the NEC requirements for motors in Section 430.14.

a. true
b. false

77. NEC Section ________ states that a circuit directory must be completed legibly so that it can be used by another electrician or understood by emergency personnel.

78. Which of the following NEC articles covers EMT installation requirements?

a. 90
b. 358
c. 511
d. 705

79. For 12 AWG nonmetallic sheathed cable, size ________ ground wire applies, while 8 AWG nonmetallic sheathed cable takes size ________ ground wire.

80. It is sometimes permissible to install any receptacle in a kitchen that lacks GFCI protection.

a. true
b. false

81. The door to an indoor transformer vault should be fire rated to ________ hours.
  a. 2
  b. 3
  c. 5
  d. 12

82. According to NEC Section 240.51, Edison plug fuses are limited to
  a. 10 amperes.
  b. 20 amperes.
  c. 25 amperes.
  d. 30 amperes.

83. Wire-binding screws or studs and nuts with upturned lugs or the equivalent can connect ________ AWG or smaller conductors.
  a. 6
  b. 8
  c. 10
  d. 12

84. Surge arresters less than 1 kV are also known as Type 1 ______________-______________ devices, while ____________________________ suppressors are known as Type 2 and Type 3 SPDs.

85. NEC Article 330 addresses Type AC cable, while NEC Article 320 addresses Type MC.
  a. true
  b. false

86. A flexible cord that supplies a room air conditioner should be no longer than ________ ft. for a nominal 120V rating or ________ ft. for a nominal 208 or 240V rating.
  a. 12, 6
  b. 10, 6
  c. 10, 5
  d. 12, 8

87. Motors must have more than one disconnecting means.
  a. true
  b. false

88. When voltage across a resistance doubles and the resistance is halved, the current increases.
  a. true
  b. false

89. The term ________________ indicates a chassis is isolated from a system ground.

90. Aluminum cannot be used for grounding electrodes.
  a. true
  b. false

91. NEC Article 760.3 states that NEC Article ________ is also referenced when defining acceptable work for fire alarm systems.
  a. 200
  b. 300
  c. 400
  d. 500

92. Wire mold is commonly used to safeguard wires inside a finished commercial building.
  a. true
  b. false

93. To classify as a nursing home, an operation must serve at least ________ residents.

a. 4
b. 12
c. 25
d. 50

94. NEC Section ________ outlines the requirements for sizing pull boxes.

95. Projectors in a movie theater must be mounted 5 in. apart.

a. true
b. false

96. Per NEC 210.62, receptacles above a show window must be within ________ of the top of the window.

a. 6 in.
b. 12 in.
c. 18 in.
d. 36 in.

97. Kitchen sinks require lights overhead.

a. true
b. false

98. Demand factors for elevator installations are based on a duty cycle of

a. 10%.
b. 20%.
c. 25%.
d. 50%.

99. Appliance calculations appear in NEC Sections ________ through ________.

100. A circuit that cannot cause a fire in a hazardous condition is considered a(n) ____________ ____________ system.

a. hermetically sealed
b. intrinsically safe
c. closed-loop
d. open-loop

101. The connection of electrical appliances is covered in NEC Article ________.

102. Which of the following organizations sponsors the NEC?

a. National Engineering Corps
b. National Fire Protection Association
c. Federal Union of Electrical Workers
d. Occupational Safety and Health Administration

103. The conductors of a multiwire branch circuit must arise from the same panel.

a. true
b. false

104. Per Section ________, to make overall calculations, correct NEC rounds ampacity calculations downward.

105. Which of the following NEC articles covers motors, motor circuits, and controllers with hermetic refrigerant?

a. 430
b. 440
c. 445
d. 490

106. The acronym NFPA represents which of the following?
a. National Federation of Professional Architects
b. Neighborhood Fire Protection Alert
c. National Fire Prevention Association
d. National Fire Prevention Agency

107. A receptacle on porcelain requires no GFCI protection.
a. true
b. false

108. NEC Article ________ allows extension cord use in wet areas.

109. With some exceptions, NEC requires panelboard circuit directories.
a. true
b. false

110. Feeder cable installation requirements appear in NEC Article ________.

111. Which of the following is NOT a conductor?
a. air
b. silver
c. copper
d. water
e. gold

112. NEC renders junction accessibility optional.
a. true
b. false

113. Per NEC Section ________, a flexible cord may be used without a grounding conductor or termination when listed and marked as double insulated.

114. A capacitor operating at 480V must discharge to 50V or less within ____________ of disconnection.
a. 1 second
b. 10 seconds
c. 1 minute
d. 1 hour

115. An electric baseboard heater cannot reside below a receptacle outlet.
a. true
b. false

116. "An apparatus designed to control and organize unused lengths of cable or cord at electrified truck parking spaces" best defines a
a. cord connector.
b. parking space disconnecting means.
c. cable management system.
d. truck coupler.

117. According to NEC ________, a light switch is needed at each doorway.

118. When installing an autotransformer, it is important to insert an OCPD in series with the shunt winding.
a. true
b. false

119. Per NEC Article 410, drop ceiling tiles can support luminaires when those the luminaires are
a. less than 5 pounds.
b. smaller than 3 in.
c. in dry areas.
d. none of the above

120. NEC Article ________, Part ________ covers computed loads and demand factors.

121. Bulbs with 750 total watts have ________ amps at 120V.

a. 8.45
b. 7.3
c. 6.25
d. 9.0

122. NEC Section 250.28(A) covers the type of material electricians must use for grounding and bonding.

a. true
b. false

123. The requirements for burying 240V conductors appear in NEC Table

a. 220.42.
b. 300.5.
c. 310.73.
d. 430.7.

124. The connection between a grounded circuit conductor and an equipment grounding conductor at a separately derived system is the

a. equipment bonding jumper.
b. system bonding jumper.
c. grounding electrode conductor.
d. ground fault.

125. A set of lights can have only a limited number of switches.

a. true
b. false

126. In a multiphase system in which one phase is used as a single phase source, the ____________ conductor serves as ground.

127. Current is measured in units known as ____________.

128. Disconnects and appliance switches must indicate their on and off positions.

a. true
b. false

129. In a circuit with 2 amperes of current and 12V, ________ watts of power dissipate.

a. 6
b. 8
c. 12
d. 24

130. NEC Section ________ covers the routing and sizing of grounding conductors.

131. A ground ring around a building or structure, in direct contact with the Earth and consisting of at least ________ ft. of bare copper not smaller than # ________ AWG, is an allowable electrode.

a. 10, 4
b. 20, 4
c. 25, 4/0
d. 20, 2

132. Where does the codebook outline definitions for AC and refrigeration equipment?

a. Article 90
b. Article 100
c. Article 250
d. Article 440

133. Per NEC Article 280, a surge protector can replace a bonding conductor.

a. true

b. false

134. Guidance for employing a portable AC generator appears in NEC Section

a. 250.34.

b. 276.4.

c. 335.11.

d. 560.52.

135. NEC Section ________ recommends limiting the voltage drops or feeders.

136. According to NEC Table C.1 in Annex C, 38 2/0 kcmil XHHW-2 conductors can be pulled into 4 in. trade size conduit.

a. true

b. false

137. In electrical terms, flexible metal conduit is known as ______________.

138. In a circuit with a 10V source and 2 ohms of resistance, a fuse would be ________ amperes.

139. Doorways require light switches.

a. true

b. false

140. NEC Table 220.11 helps electricians calculate lighting loads in restaurant dining areas.

a. true

b. false

141. Electricians should staple AC cable ________ from the box.

a. 4 in.

b. 6 in.

c. 8 in.

d. 12 in.

142. NEC Article 700 oand its subsidiaries covers Class III locations only.

a. true

b. false

143. The fill requirements for conduit installed outdoors appears in NEC Article

a. 225.

b. 380.

c. 770.

d. 821.

144. Gas hot water heaters require electricity.

a. true

b. false

145. Fuse and breaker requirements are covered by the same NEC article but different sections.

a. true

b. false

146. When installing parallel conductors, electricians should consult NEC Section ________.

147. Wire running inside island cabinets require no protection.

a. true

b. false

148. A 12 AWG wire requires a volume of ________ cubic inches.

a. 2.0
b. 2.25
c. 3.0
d. 3.25

149. What NEC Section discusses wiring of elevators?

a. 620.14
b. 620.21(A)
c. 250.12
d. 570.74

150. A box requires how much outer jacket be left inside the box?

a. 0 in.
b. $\frac{1}{3}$ in.
c. $\frac{1}{2}$ to $\frac{1}{4}$ in.
d. none of the above

151. NEC considers the Earth effective and permissible as a bonding jumper between two grounding rods.

a. true
b. false

152. A switch should be ________ off the floor.

a. 4 in.
b. 8 in.
c. 12 in.
d. 48 in.

153. NEC Article ________ outlines the requirements for installing raceways under floors.

154. It is acceptable electrical practice to install lights in shower stalls.

a. true
b. false

155. When electricians confront rooms in offices that are designed to be computer rooms but do not function as such, those electricians should consult which NEC section?

a. 338.12
b. 514.4
c. 392.5
d. 820.54

156. In a patient bed location in a critical care area, all branch circuits from the normal system should arise from one panelboard.

a. true
b. false

157. A receptacle or switch accepting 14-2 counts as ________ cubic inches.

158. According to NEC Section ________, a disconnect is not required when the breaker in a panel can be locked in the "off" position.

159. With a GFCI protected circuit, a switch rated for dry use only can function in a wet area.

a. true
b. false

160. What is the minimum diameter for a round manhole?

a. 12 in.
b. 26 in.
c. 36 in.
d. 48 in.

161. Electrical requirements for unfinished basements appear in NEC Section ________.

162. When connecting grounding and bonding equipment, connection devices or fittings that depend solely on solder are permitted.
a. true
b. false

163. Conductors, according to NEC 110.5, are assumed to be
a. copper.
b. aluminum.
c. brass.
d. none of the above

164. For NEC purposes, a garage-door opener is considered an appliance.
a. true
b. false

165. Exposed ballasts may be in contact with flammable materials.
a. true
b. false

166. After determining feeder capacity, electricians should choose panelboards using which NEC section?
a. 314.7
b. 547.2
c. 408.13
d. 705.13

167. Treating soil with salts and chemicals to enhance grounding ability is acceptable practice.
a. true
b. false

168. According to NEC Section ________, the zone, which equals the width and depth of equipment from floor to 6' above the equipment, should be kept clear of all piping, ducts, or other equipment unrelated to the electrical installation.

169. A UL-approved and inspected device requires no reinspection.
a. true
b. false

170. Microwaves and garbage disposals should operate on the same circuit.
a. true
b. false

171. The topic of NEC Table 450.3 is which of the following?
a. transformer ventilation
b. transformer load-side surge suppression
c. transformer supply-side OCPD
d. none of the above

172. Restrictions on the use of plastic junction or device boxes appear in NEC Article ________.

173. When installing supplementary electrodes, electricians must consult which of the following NEC sections?
a. 110.31(C)
b. 250.104(A)
c. 342.5(A)(1)
d. all of the above

174. Wire that goes to a device must extend 8 inches past the expected termination point.
a. true
b. false

175. According to NEC Section ________, all the phase wires in parallel conductors must be the same length, type, and size.

176. NEC 250.30(A) requires an Earth connection on the load side of the grounding point of a separately derived system.
a. true
b. false

177. A grounding electrode conductor must be installed in one continuous length according to NEC Section ________.

178. The metal yoke of a receptacle is common to the screw terminal for ground unless the receptacle is an isolated ground receptacle.
a. true
b. false

179. Power wires should be ________' from an attic access hole.
a. 2
b. 4
c. 6
d. 8

180. The requirements for thermal ballast protection appear in NEC Section ________.

181. A residence with four bedrooms on a second floor with one hall area, one bedroom on a first floor with two hall areas, and one basement would require how many smoke detectors?
a. three
b. five
c. seven
d. nine

182. NEC Article ________ covers the installation of lighting fixture and supports.

183. Which of the following NEC sections covers fluorescent ballasts that are installed indoors and require internal thermal protection?
a. 410.7(E)(1)
b. 457.4(A)(3)
c. 477.56(D)(2)
d. none of the above

184. NEC 410.8 identifies the special requirements of installing lights in a clothes closet.
a. true
b. false

185. Panelboards equipped with snap switches rated at 10A or less have an overcurrent protection of 200A or less.
a. true
b. false

186. NEC Section ________ states that when equipment is marked in the maximum fuse size, breakers are acceptable.
a. 110.3(B)
b. 233.4(C)
c. 410.7
d. 543.2

187. Emergency circuits can sometimes run in the same box as normal wiring.
a. true
b. false

188. To prevent electrical noise in a building's framework from affecting sensitive equipment, grounds should be bonded.
a. true
b. false

189. If, once drywall is installed, the lip of a box is recessed too much, the electrician may mount the switch or outlet yoke directly to the wall.

a. true

b. false

190. NEC's purpose appears in which section?

a. Appendix C

b. Appendix A

c. 110.1(A)

d. 90.1(A)

191. In reference to switchboards and panelboards, the phrase *condition one* means ______________.

192. NEC Section _________ contains the standard ampere rating for fuses and nonadjustable circuit breakers.

193. A grounded conductor termination bar should not be insulated from the enclosure of a park trailer distribution panelboard.

a. true

b. false

194. Examples of dedicated circuits in a residence include the ____________________, ____________________, ____________________, and ____________________.

195. According to NEC Section _________, a disconnecting means must be at least 115% of the sum of all loads in equipment.

a. 440.12(A)(1)

b. 440.12(B)(2)

c. a and b

d. none of the above

196. NEC Section _________ discusses overload protection for motor compressor appliances.

a. 230.42

b. 440.52

c. 560.32

d. none of the above

197. A heavy-duty lampholder, per NEC 210.21(A) must be rated at 600W or greater.

a. true

b. false

198. The bonding jumper insures a connection between metal parts and the earth.

a. true

b. false

199. When an appliance is unmarked with a breaker requirement, the maximum current rating is omitted.

a. true

b. false

200. According NEC Section, earth is not to be the only equipment grounding conductor.

a. 708.55(A)

b. 645.2(B)(1)

c. 310.3(C)

d. 250.4(A)(5)

201. NEC Table _________ contains information on the volumes of different conductors.

202. A fixed electric space heater is considered an intermittent or a noncontinuous load when calculating branch circuit ampacities.

a. true

b. false

203. The definition of the term *bundled* appears in which NEC article?

a. 310
b. 100
c. 520
d. 334

204. The requirements for sizing main bonding jumpers appear in NEC Article ________.

205. The term ____________ means "acceptable to the AHJ."

a. *listed*
b. *tested*
c. *approved*
d. *guaranteed*

206. Where a multiwire circuit's disconnecting device opens all ungrounded conductors simultaneously, a multiwire branch circuit is allowed in a Class ________, Division ________, location.

a. 2, 2
b. 2, 1
c. 1, 2
d. 1, 1

207. NEC Section ________ requires two independent equipment grounding paths.

208. Non-current-carrying conductive materials enclosing electrical conductors or equipment are covered by NEC Section ________.

209. All kitchen lighting fixtures must be on the kitchen circuit.

a. true
b. false

210. Which NEC section directs electricians to avoid regrounding the neutral at building supplied by a feeder from the service equipment?

a. 250.32(B)
b. 253.32(B)
c. 255.42(A)
d. all of the above

211. The ____________ of an electrical installation can order a building or residence disconnected from its power supply.

212. NEC Section ________ requires all nonconductive coatings be removed before grounding or bonding.

213. Recessing a panelboard into a flammable wall is acceptable practice.

a. true
b. false

214. A site-isolating device installed at a distribution point serving two or more agricultural buildings or structures should be permanently marked ____________ the operating handle to identify its purpose.

a. on the face of
b. immediately adjacent to
c. within $6\frac{1}{2}$ ft. of
d. a and b

215. Separately derived systems are covered under NEC Section ________.

216. All independent grounded neutral systems must be grounded according to NEC Sections ________ and ________.

217. The neutral current in a 208Y/120V four-wire system when the phase currents are 20, 40, and 60 amperes through phases A, B, and C, respectively, is 32 amperes.

a. true
b. false

218. "The current, in amperes, that a conductor can carry continuously under the conditions of use without exceeding its temperature rating" is known as

a. power rating.
b. ampacity.
c. temperature rated current.
d. continuous load current.

219. When a premises' wiring system that is supplied by a grounded AC service must have a grounding electrode attached to a grounded service conductor, the electrician should consult which of the following NEC sections?

a. 680.6(B)(2) through (B)(4)
b. 250.24(A)(1) through (A)(5)
c. a and b
d. none of the above

220. NEC recall is the most important aspect of learning the electrical trade.

a. true
b. false

221. The symbol   represents the entity known as the ________________.

222. NEC Section ________ requires unspliced main bonding jumpers.

223. In overhead service installations, when no local regulations apply, the service head must comply with NEC 314.15.

a. true
b. false

224. The general rule that a feeder conductor must have a minimum allowable ampacity of 125% of the noncontinuous load plus 125% of the continuous load fails to consider conductor

a. diameter.
b. radius.
c. area.
d. temperature.

225. A gas stove can operate on a kitchen counter circuit.

a. true
b. false

226. NEC requirements guarantee top workmanship and the safest installations.

a. true
b. false

227. Ambient conduit temperature rating above a roofline depends on how far the conduit is above the roof.

a. true
b. false

228. When using equipment bonding jumpers, electricians should reference NEC Section ________.

229. In a cupboard, M/C cable should replace Romex to
- **a.** extend cupboard space.
- **b.** increase amperage.
- **c.** safeguard the wire.
- **d.** all of the above

230. A bathroom exhaust fan requires GFCI protection.
- **a.** true
- **b.** false

231. Per NEC Article ________, an underground location is considered a wet environment.

232. Electrode-type boilers operating at over 600 volts, nominal, should be supplied only from a three-phase, ____________ system or from isolating transformers arranged to provide such a system.
- **a.** four-wire, solidly grounded wye
- **b.** corner-grounded delta
- **c.** a and b
- **d.** none of the above

233. In a three-phase panelboard, the phases must be arranged from left to right or from top to bottom as seen from the front.
- **a.** true
- **b.** false

234. The electrical noise from one conductor that affects the signal in another conductor is known as
- **a.** crosstalk.
- **b.** amplification.
- **c.** coupling.
- **d.** brownian noise.

235. A circuit breaker listed and marked as HID may be used to switch 120 and 277V fluorescent lighting circuits.
- **a.** true
- **b.** false

236. Electricians may use ____________ to strap wire to a three-gang box.

237. A panel should be ________ in. off the floor.

238. A(n) ____________ niche light requires surrounding water to cool its bulb during operation.

239. Smoke detectors cannot reside on a separate circuit because they are routinely tested.
- **a.** true
- **b.** false

240. Per NEC Section ________, electricians may hang secondary supports from primary supports when running new work over old.
- **a.** 90.1
- **b.** 109.12
- **c.** 300.11
- **d.** 415.3

241. In theaters, motion picture and television studios, and similar locations, the neutral conductor of feeders supplying systems that use or may use both phase-control and sine wave dimmers should not be considered as current-carrying.
- **a.** true
- **b.** false

242. The tightness of a turn in wire or conduit is called

a. curve constant.
b. bend radius.
c. turn radius.
d. 90° angle.

243. Edison-base plug fuses should be classified as not over ________ volts and ________ amperes.

a. 277, 50
b. 240, 45
c. 220, 40
d. 125, 30

244. Splices and taps in a wireway must be inaccessible to ensure safety.

a. true
b. false

245. According to NEC Table 5A, the diameter of a bare number 4 AWG conductor is

a. 0.175 in.
b. 0.1875 in.
c. 0.213 in.
d. 0.225 in.

246. Each patient bed area must be on a dedicated branch circuit in a healthcare facility.

a. true
b. false

247. The organization responsible for developing and maintaining standard symbology for prints and schematics is

a. ANSI.
b. NFPA.
c. NEC.
d. SEC.

248. When classifying locations in spray applications, dipping, and coating processes, the interior of any open or closed container of a flammable liquid should be considered a Class ________, Division ________, or Class ________, Zone ________, location.

249. Which of the following items is not allowed to be connected to the supply side of a service disconnect?

a. current limiting device
b. photovoltaic systems
c. instrumentation transformers
d. all of the above

250. A cablebus is not considered a wiring method for service entrance installation.

a. true
b. false

Answers and Explanations

▶ Pretest Review

1. **b.** According to NEC Article 100, a device that opens and closes a circuit by nonautomatic means automatically closes for a predetermined overcurrent.

2. **c.** Annex C of NEC covers conduit and tubing fill for conductors and fixture wire fill.

3. **a.** NEMA is the acronym for National Electrical Manufacturers Association.

4. A receptacle with NEMA 5-20R on it has an amperes rating of 20, will take a NEMA-5-15P or NEMA-5-20P plug, and has a voltage rating of 125.

5. **a.** NEC Tables 310-10 through 310-21 cover the sizes of copper and aluminum wire, as well as the coating of wire that has fewer than three current-carrying wires in a cable or conduit or that is buried in the earth.

6. The fill of metal boxes appears in NEC Tables 314.16 A and B.

7. **c.** NEC Article 358 covers EMT.

8. NEC Articles 342, 344, and 358 outline the installation requirements of RMC, IMC, and EMT.

9. This is the symbol for a floor receptacle box with a duplex receptacle.

10. A three-way switch is depicted by this symbol.

11. **b.** This symbol represents a device that is surface mounted.

12. A door bottom is depicted by this symbol.

13. This symbol represents a panelboard.

14. This symbol means the end of a branch circuit to panelboard.

15. **d.** This symbol represents a switch with a dimmer.

16. In NEC 2008, Table 370.16A shows, in two columns (millimeters and inches), how many wires in trade size you can place in a box.

17. NEC Chapter 9, Table 4 outlines trade size of conduit as well as raceway size.

18. NEC Article 386 covers metallic or metal raceway; NEC Article 388 covers PVC pipe.

19. **c.** NEC Article 408 covers panel board application.

20. **c.** A 12-in. counter requires a receptacle because it holds small appliances.

21. According to new codes, ARC fault protection is now needed in many other parts of the house. See NEC 210.12, 210.12A, 210.12B, 440.65, and 550.25.

22. **a.** One receptacle strap is required by NEC; two are suggested. NEC 210.52(H) defines as a hallway.

23. **d.** Definitions are the subject of NEC Article 100.

24. The main ground on a service is a ground rod.

25. NEC Section 250-28 requires a main bonding jumper and a system bonding jumper.

26. **a.** NEC Section 250-28A provides information on what material can be used for a main bonding jumper.

27. **d.** A bonding screw is green.

28. **b.** A bonding screw is unneeded in a subpanel.

29. According to NEC Section 250-28D, the three methods of sizing bonding jumpers are (1) bonding jumper is sized not less than 12.5% of the largest phase conductor, (2) based on the largest ungrounded service conductor, and (3) based on the largest sum of the service conductors.

30. **b.** NEC Table 250.66 covers grounding electrode conductor sizing.

31. NEC Tables Annex C1 to C12 identify both wire sizing and the temperature wire can handle in amps.

32. According to NEC, a grounding system is connected to the ground or to a conductive body that extends the ground connection.

33. **b.** NEC Section 250-122 is dedicated to the sizes of grounding conductors.

34. In NEC terms, a circuit breaker is a device designed to trip when an overcurrent is detected.

35. **a.** NEC Section 300-13 provides information on mechanical and electrical continuity conductors.

36. A bonding jumper system is the subject of NEC Section 250-2D.

37. **c.** NEC Section 680-20 discusses electrical installations at permanently installed pools.

38. **a.** NEC Section 250-118 covers equipment grounding conductors.

39. **c.** NEC Section 250.119 1–14 identifies equipment grounding conductors.

40. **a.** NEC Section 250-126 covers ground terminal identification.

41. Bonding is the establishment of reliable conductor connection between metal parts.

42. NEC Section 250-134 discusses equipment fastened or connected by permanent wiring methods.

43. **d.** Equipment grounding conductor connections are discussed in NEC Section 250-130.

44. **d.** According to NEC Section 250-118, a grounded conductor must run inside flexible conduit.

45. NEC Section 250.118(5) oversees bonding of flexible type conduit.

46. NEC Article 392 covers cable trays.

47. A bonding jumper should be installed inside or outside raceway within 6 ft. of attachment (NEC 250.118).

48. **b.** Threaded pipe is appropriate for bonding (NEC 250.92B).

49. **a.** Shoulder fittings with locknuts are allowable as bonding devices (NEC 250.92B).

50. NEC Section 250-66 covers the sizes of alternating current grounding electrode conductors.

51. **b.** Equipment bonding jumpers are covered in NEC Section 250-102.

52. Electricians can use sheet metal screws for bonding when those screws are green and have hexagonal heads (NEC 250.8), but it is generally considered unwise.

53. NEC Article 230 covers service drops.

54. **b.** NEC does not approve trees as service conductors.

55. **b.** No disconnect is required between meter base and panel when a service entrance wire is under at least 2 in. of concrete.

56. **a.** Above sidewalks accessible by pedestrians, a service drop not over 600V must be at least 10 ft. high.

57. A 200-amp service crossing residential property requires a 12 ft. service drop distance.

58. **d.** The required clearance for a 400-amp service drop across a major roadway is 18 ft.

59. NEC Article 230.9 covers clearances on buildings.

60. According to NEC 230.9(A), an unsheathed service conductor must be 3 ft. from a window that opens.

61. c. NEC Article 250 covers grounding.

62. a. System and circuit grounding involves taking alternating current to ground.

63. Equipment grounding exposes current-carrying parts to a grounding system.

64. Insulated housings and grounds on the ends of plugs are two ways to protect small power tools.

65. b. Run ground wire from panel to copper water pipe then jump the water meter using grounding clamps.

66. To ground a subpanel in the same building, run a four-wire system from main panel to subpanel.

67. To ground a subpanel in another building, run bare copper wire from subpanel to ground rod inside or outside.

68. To be grounded is to be connected to ground or to a conductive body that extends the grounding connection.

69. A grounded conductor is a system or circuit that is intentionally grounded.

70. A grounding conductor is a conductor or circuit of a wiring system to a grounding electrode (ground rod).

71. An electron is the fundamental particle of electricity.

72. b. Current is fatal to people, not voltage.

73. d. Voltage level, amount of protective clothing, and moisture on skin or outerwear all impact a ground connection.

74. a. Grounding reduces the voltage that might cause a fatal electrical shock by allowing excess current to flow to the ground.

75. Ground and arc are two types of fault in electrical work.

76. A 400-amp service requires two ground rods 6 ft. apart.

77. Two outside ground wires are needed on a 400-amp service.

78.

B High Leg Orange
240V
A
C
Red
240V Black
Neutral 120V White

79.

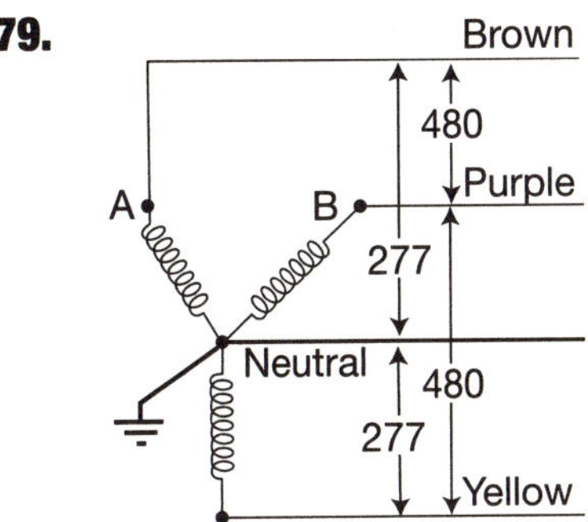

80. a. NEC allows electricians to ground gas pipes to existing water pipes when those water pipes are grounded.

81. Building steel is permitted in an electro grounding system only when the steel is connected to Earth ground.

82. The size of the largest ungrounded service entrance conductor determines what size ground wire is used on a service.

83. a. A grounding conductor is green.

84. An isolated grounded reciprocal is a separate ground from a device directly to a panel.

85. Exothermic welding involves the fusion of metal materials or copper to rebar.

86. a. Electricians may use rebar as a grounding means.

87. A direct burial clamp is needed to properly connect number 4 copper to a ground rod.

88. According to NEC Section 250-8, five methods of making connections include

a. listed pressure connectors

b. terminal bars

c. pressure connectors listed as ground and bonding equipment

d. exothermic welding

e. machine screw-type fasteners that engage not less than two threads or are screwed with nuts

89. **b.** Relative to underground service, overhead electrical service is simpler to troubleshoot and repair and is the easier option when running electricity across roadways, waterways, and right-of-ways.

90. In NEC, Table 250.66 provides the sizing of grounding electrode conductors.

91. To make a proper grounding system using three 200-amp disconnects, use a ground wire of sufficient size to go from panel to panel with a lug, screw, or ground bar, if provided.

92. The smallest grounding conductor permitted is 6 AWG or larger.

93. **a.** The way to ground a transformer in a commercial building is to place a grounding rod outside or beneath the building.

94. Two hots are needed to operate the stove neutral. The timers and clocks can serve as ground for the tripping mechanism.

95. With dryers, two hots are needed to operate the heater neutral. Timers can act as ground for the tripping mechanism.

96. **d.** All of these can operate safely without a four-wire system. For example, air conditioning needs only 220 and a ground to complete the circuit.

97. **c.** NEC Article 250.114 outlines which air-conditioning systems require no grounding.

98. Isolated systems are used in healthcare facilities due to explosive gases.

99. **b.** Grounded conductors belong on green screws on reciprocals.

100. On a 200-amp service, the electrician should use 2/0 or 4/0 grounded conductor in the service cable.

101. When using 500 MCM, 350 MCM or 400 MCM size ground wire is appropriate on a 400-amp service.

102. High impedance grounding is used for grounding radio towers (see NEC Article 250.20[E]).

103. To test conduit for proper grounding, check ohms or use a 120V receptacle and touch to ground to check for voltage.

104. Per NEC 90.4, 510.2, and various local codes, occupancy is permitted in the event of a code violation as long as the occupancy remains unchanged and no hazardous condition exists.

105. **d.** Per NEC Table 402.3, FFH-2 fixture wiring is usable up to 75°C.

106. **c.** Bonding requirements for healthcare facilities are discussed in NEC Article 517.

107. **b.** Per NEC 360.12, FMT is disallowed from battery storage areas.

108. **b.** NEC calls a circuit that cannot cause a fire in a hazardous condition an intrinsically safe system.

109. NEC uses the phrase, "Authority Having Jurisdiction (AHJ)" to refer to local governing of construction procedures and standards.

110. **c.** A dedicated circuit is designed and created for one load.

111. c. Chapter 9 tables specify conduit fill requirements for fiber optic cabling.

112. b. Chapter 3 of the NEC covers the use of supplied fixture wiring, which is more representative of terminations, not wiring methods.

113. f. The text from NEC 90.1(A) states that NEC's purpose is the "practical safeguarding of people and property from hazards arising from the use of electricity," which includes all choices.

114. Resistive, impedance, and skin effect are the heating systems called out in NEC Article 426.

115. Per NEC 408.3, phases must arrange in a particular order: A, B, C, from left to right or from top to bottom as seen from the front.

116. b. Per NEC 450.4, when installing an autotransformer an OCPD need not be installed in series with the shunt winding.

117. b. NEC Section 90.1 C describes NEC overall as a minimum standard.

118. The NEC Article 100 statement, "All circuit conductors between the service equipment, the source of a separately derived system, or other power supply source and the final branch-circuit overcurrent device," is the definition for *feeder*.

119. NEC Article 314.3 discusses restrictions on nonmetallic boxes.

120. c. NEC Chapters 1–4, which govern basic practices, remain in effect in hazardous areas, except as modified by specific requirements outlined in NEC Chapter 5.

121. d. Class III hazardous areas may contain airborne particulate matter, like sawdust, that can be ignited by a spark.

122. a. Per NEC 504.10(B), general-purpose enclosures are NEC approved for intrinsically safe apparatuses. Intrinsically safe is otherwise safe without an enclosure, so the enclosure offers no added protection. The enclosure is, however, required for other types of protection.

123. b. Fuses on the load side do not protect transformers. Per NEC 450.3, when a transformer is enclosed, install OCPD at the transformer's primary.

124. b. Per NEC Section 100, ampacity is "the current, in amperes, that a conductor can carry continuously under the conditions of use without exceeding its temperature rating."

125. NEC Article 310.8 lists appropriate conductor insulation types for wet locations. NEC 310.9 and 310.13 provide more specific information on the selected type.

126. a. NEC 110.23 states that, "Unused current transformers associated with potentially energized circuits shall be short circuited."

127. NEC 220.16 through 220.20 cover appliance calculations.

128. Typically, transformers come in liquid-filled and dry versions (NEC Article 450, Sections 450.21–28).

129. Use NEC Chapter 9, Table 5, to locate wire type and size to be installed. Repeat for each conductor, keeping a running total. Derate by 40% to get the total raceway area. Refer to Chapter 9, Table 4, for the type of raceway to be used. Find a value for area larger than the raceway area calculated previously.

130. a. Depending on the location and environment, 3 ft. may be insufficient clearance. While grouping apparatuses tightly may save space, repair and maneuvering options are limited should a device fail.

131. **a.** NEC 225.18 discusses "clearance(s) from ground." It is a good idea to exceed the requirement by a foot or two given that the clearances NEC outlines reflect a minimum.

132. **b.** Per NEC 230.24, which discusses service drop and other clearances, "Service drop conductors shall not be readily accessible."

133. **a.** Electricians may use extension cords in wet areas when those cords are rated for use in wet areas, per NEC Article 400.

134. **b.** Liquid-filled transformers have different requirements based on the dielectric fluid they use. Reference NEC 450.23–26 as needed.

135. **d.** Deicing systems are discussed in NEC Article 426, not NEC Article 526.

136. **a.** Switches or circuit breakers used as a means of disconnect must indicate switch status, per NEC 422.35.

137. **b.** Per NEC 408.3, the high-voltage leg must be identified as the "B" leg.

138. **a.** Per NEC 210.11, loading per unit area must be evenly distributed among branch circuits.

139. **b.** NEC Article 312.3 disallows installation of recessed boxes in combustible walls and specifies the relief in noncombustible walls.

140. **b.** NEC Article 314.4 requires all metal boxes to be grounded.

141. **b.** A device's nameplate tells what size breaker to use, per NEC 422.11.

142. **b.** NEC Table 402.5 defines ampacity requirements for fixture wires.

143. **b.** NEC Article 430 Part III sets out requirements for motor protection. Protecting a motor and protecting a motor circuit are two very different things.

144. **d.** Ceiling tiles are never appropriate device support.

145. NEC Table 300.5 provides the requirements for buried cable up to 600V.

146. NEC Article 320 presents requirements for type AC.

147. **a.** NEC considers garage door openers appliances (NEC 422.31).

148. **b.** Per NEC 422.31, appliances with motors over 0.125 hp must have a separate means of disconnect.

149. **a.** Per NEC 300.11, hanging a second support from a primary support when running new over old is acceptable and a common practice.

150. **c.** Although NEC 450.46 only requires a drain in the event of transformer action exceeding 100KW, installing one is good idea practice so as to contain and clean dielectric spills.

151. **b.** A TVSS is designed not to provide protection against lightning but to guard against known and predictable anomalies, including short-duration spikes of high voltage.

152. **a.** According to the NEC, equipment must be installed to the manufacturer's instructions and specifications. If this is not done, the listing of the apparatus is void and the installation violates NEC.

153. **b.** Bonding of all cable tray sections is required per NEC 392.7(B).

154. **d.** A receptacle used with aluminum wiring must be marked CO/ALR.

155. Per NEC Table 220.13, use 100% of the first 10 KW and 50% of the demand above 10 KW. In this case, the demand factor would be 10,000 plus half the remaining 10,000 or 15,000W.

156. **c.** Per NEC 225.34, a maximum of six disconnects are allowed per service entrance.

157. **c.** In a given NEC section, Subsection (2) contains terms and definitions.

158. **a.** In the right conditions, even overcurrent insufficient to trip a breaker can prove hazardous.

159. **e.** According to the NEC 2008, all of these areas must be AFCI protected.

160. In a multiphase system in which one phase is used as a single, phase source, the neutral conductor serves as ground, per NEC 250.26(5) and (2).

161. Per NEC Article 100, the phrase *in sight* means visible and within 50 ft.

162. **d.** The NEC is also known as NFPA 70.

163. As part of an FPN after the definition of "Authority Having Jurisdiction," NEC Article 100 allows property owners to act as their own AHJs.

164. Per Table 220.84, a 21-unit building would be rated at 32% and any occupancy over 62 units rated at 23%.

165. **d.** All these statements are true, per NEC 250.68(B).

166. NEC Article 100 defines bonding as "The permanent joining of metallic parts to form an electrically conductive path that ensures electrical continuity and the capacity to conduct safely any current likely to be imposed."

167. **a.** NEC Table 250.66 applies to the derived conductors of separately derived AC systems (Note 2 subscript a of the table).

168. **d.** Per NEC 250.52(A)(4), ground ring encircling the building or structure, in direct contact with the earth consisting of at least 20 ft. of bare copper not smaller than #2 AWG, shall be permitted as an electrode.

169. **d.** Per NEC 250.28(B), where a main bonding jumper or a system bonding jumper is a screw only, the screw shall be identified with a green finish that shall be visible with the screw installed.

170. **b.** The demand factors of NEC Table 220.42 do not apply in situations when lighting is run as a concurrent load, such as in a restaurant where lights are on all the time.

171. **c.** Outdoor lighting installations exceeding 120V are subject to the clearance requirements outlined in NEC 225.7.

172. **c.** Per NEC 225.14, open conductors belonging to separate circuits must be separated by 4 in. (100 mm).

173. **a.** NEC 225.37 requires a panelboard circuit directory, although there are exceptions.

174. **b.** NEC 230.10 disallows vegetation as a support type.

175. **a.** NEC 230.40 states that, "Every service drop or lateral shall supply only one set of service entrance conductors."

▶ Introduction Review

1. In electric circuits, a load usually converts electrical energy to heat, magnetism, and light, all forms of energy.

2. Using Ohm's law, the fuse value would be 10V/2 ohms or 5 amperes. Thus, a 5-amp fuse would theoretically suffice. However, if the voltage were to increase slightly and harmlessly, the fuse would blow for no reason. Therefore, fuse selection is not as precise as it would seem and should be based on the nature of the load, the consequences of overcurrent, and other, lesser factors.

3. **b.** The term *electricity* is defined as the imbalance in the negative charges, or electrons, between two points.

4. In DC, electrons flow in one direction only. In AC, electrons "saw" back and forth in the load. AC is easier to manufacture and is most often delivered and used in a residence, business, or factory.

5. **e.** Air does not conduct electricity. Gold, silver, copper, and water all do, to varying degrees.

6. Current is measured in amperes, named for French scientist, André Marie Ampére.

7. **b.** Arcing, the phenomenon in which electrons jump across insulators to positive terminals, can occur through air, paper, rubber, or any other insulator.

8. Given that power equals voltage times resistance, power in this example is 24 watts (2 amps × 12V).

9. **a.** All electrical equipment and supplies should be considered active or live until determined otherwise.

10. **d.** Voltages of 120 or more, the minimum voltage typically supplied by utility companies, are considered high.

11. **c.** If the voltage is doubled and the resistance is halved, the current remains the same, because the doubling offsets the halving.

12. **b.** All voltages in a circuit are referenced to earth ground, or 0V, through a green wire.

13. **b.** In a parallel circuit, each load is connected in a separate path. In a circuit schematic, the branch lines are parallel.

14. Safety is the most important aspect of an electrician's job.

15. When a transformer has a voltage ratio of 8:1 and an input of 120V, the output voltage is $\frac{1}{8}$ the input, $\frac{120}{8}$ or 15V.

16. **b.** When an ammeter is connected in parallel with a load, there will probably be damage to the meter and/or circuit and possible personal injury. To measure current, the meter must be in the circuit (i.e., in series with the load it is measuring current through).

17. **a.** Motors convert electricity to rotational motion, while generators convert rotary motion to electricity.

18. **b.** The National Fire Protection Association (NFPA) sponsors the NEC.

19. **b.** In a battery, a negative terminal has a surplus of electrons, while a positive terminal has a surplus of protons.

20. **c.** Water lacks insulation qualities because it conducts electricity. As a result, it is one of the most hazardous substances in an electrical environment.

21. **a.** Current always flows from negative to positive.

22. **d.** In the standard U.S. system of wire sizes, from No. 50 through No. 4/0, the lower the number, the thicker the wire. Gauge was designed to reflect the number of wires that can fit through an opening.

23. **d.** Heart fibrillation is the major cause of death from electrical shock. The heart itself is an electrical device, one that gets its electrical signal from the brain. Problems arise when that signal or the pathway through which it travels is disrupted.

24. **b.** In wire, a smaller diameter increases resistance to flow.

25. A circuit breaker, or simply a breaker, acts as a fuse but is usually resettable. Most modern homes and other buildings have circuit breakers because they are far more convenient than fuses.

26. **c.** If you have 750 watts in bulbs, at 120V it is 6.25 amps because amperes equals watts divided by volts.

27. **a.** In electricity, resistance is generally used to convert one form of energy to another.

28. The larger the diameter of a piece of wire, the smaller its AWG value. For example, 20-gauge wire is thinner than 10-gauge wire, because wire gauge is based on fill rate (how many wires can fit into a box or conduit). Large wires take more space so fewer can be routed through (or fill) a length of conduit.

29. **b.** If you have 1,800 watts for baseboard heat, at 240V there are 7.5 amps because amps equal watts divided by volts.

30. **a.** Voltage strengthens as a negative terminal becomes crowded with electrons.

▶ Chapter 1 Review

1. **a.** Because garage space is not considered livable space, subtract its area from the total living area. The home's livable area is 28 ft. × 40 ft, or 1,120 sq. ft.

2. **b.** At minimum, NEC requires 3 watts per residential square foot to help electricians size services and determine the number of required general lighting circuits. In certain living areas, like the kitchen, more wattage may be desired for better lighting.

3. **a.** Although not required by NEC, in general any receptacles below grade, including those in finished basements, should be GFCI protected. Receptacles in dwelling accessory buildings with floors at or below grade level require GFCI protection (NEC 210.8[A][2]). Unfinished basements require GFCI-protected receptacles (NEC 210.8[A][5]).

4. **c.** Each receptacle or switch counts as 2 conductors (NEC 314.15), and 14 AWG wire conductors require 2 cu. in. each. Therefore, 2 conductors × 2 cu. in. = 4 cu. in. Add to this, 2 for the yoke and 2 for the ground for a total of 8 cu. in.

5. **a.** A grounding system designed as a main service ground takes priority over a system that provides grounding alone. While acquiring as large a grounding surface area as possible is advantageous, the grounding system must be able to support the service on its own (NEC 210.52[A][1]).

6. **a.** According to good engineering practice, a sump pump, like a refrigerator, should continue running when another component fails. A sump pump's failure manifests only when flooding occurs (NEC 680.50).

7. 14 gauge: 2, 12 gauge: 2.25, 10 gauge: 3. For this and other box fill information, please see NEC Article 314.16.

8. Breaker rating determines ampacity, or size, of a circuit.

9. According to the NEC, an outlet is defined as a point in a circuit at which current is taken (NEC Article 100). An outlet is a device that gives power to another power-drawing device.

10. Receptacles must be installed on a peninsula when the peninsula's long edge is at least 24 in. and its short edge is at least 12 in. (NEC 210.52 [C][3]).

11. First determine whether the application is residential. For residential applications, the NEC provides no guidance. The answer for nonresidential applications is based on box fill. For these applications, most electricians calculate between 1 and 1.5 amperes per receptacle.

12. **a.** Per NEC 680.90(A)(3), GFCI protection must reside on whirlpool tubs.

13. Three residential areas that require 20-amp circuits are the kitchen (NEC 210.52[A][1]), bathroom (NEC 210.11[C][1][2]), and laundry room (NEC 210.11[C][1][2]).

14. A 10 AWG wire carries 30 amps (10 × 3).

15. **b.** Due to safety and accessibility concerns, NEC requires a clearance width of 30 in. and a clearance depth of 36 in. (NEC 110.26).

16. Yes, a receptacle is required in close proximity to an air-conditioning unit. At most, it must be 25 ft. distant.

17. **a.** An attic with a furnace requires a receptacle to honor safety and sound wiring requirements.

18. **c.** Although a 15-amp breaker can protect 12AWG wiring, it is "above code" and preferable when considering expansion.

19. **c.** Along a wall space, a first outlet should be no more than 6 ft. from a door.

20. **a.** Per NEC 250.116, ground rods should attach to metal fences.

21. **b.** Per NEC, one cannot install an exhaust fan and a light in a shower area unless the installation is 8 ft. above the shower or tub rim. When elements meet this height requirement, however, they require no GFCI protection (NEC 410.10[D]) but must be listed for wet locations.

22. Among the tools for installing ground rods are sledge hammers, hammer drills, and ground rod pounders.

23. To meet good engineering practice, a receptacle should be 16 in. off the floor.

24. To meet good engineering practice, a switch should be 48 in. off the floor.

25. **a.** It is acceptable to install a lighting circuit in more than one room.

26. **a.** Receptacles fed from subpanels must have their equipment grounds connected to an equipment grounding bar that is electrically isolated from the neutral bar.

27. **b.** U F wire is typically not used for grounding purposes.

28. In NEC language, the phrase *shall be* represents a mandatory act or condition.

29. When electrical service is installed in a residence, the builder, homeowner, inspector, and boss (communication of trades) should all be satisfied with the level of work.

30. In a cupboard, M/C cable is preferred over Romex because it guards the wire against damage.

▶ Chapter 2 Review

1. **b.** Any island with a surface area is 2 sq. ft. or greater requires a receptacle (NEC 210.90[C][2]).

2. Two common types of kitchen lighting are recessed and surface.

3. An electrical service requires two ground rods that are 8 ft. long and at least 6 ft. apart (NEC Article 250). The goal is to achieve fewer than 25 ohms of resistance in the grounding system.

4. **b.** NEC requires that the movable section receptacle of a patio door be within 6 ft. of the door (NEC 210.52[A] through [Z]).

5. **a.** Each doorway in a residence requires a light switch (NEC 210.70[A][1–2]).

6. A wall space 2 ft. long must have a receptacle (NEC 210.52[A][2]).

7. A dining room receptacle requires a 12-2, 20-amp circuit (small appliance branch circuit) (NEC 210).

8. **a.** It is good practice, not NEC requirement, to place lights above kitchen sinks, although some jurisdictions may require doing so.

9. **a.** A sump pump should run on a dedicated circuit (NEC 210.20[C]).

10. An electric stove takes 6-3 size wire and a 50-amp breaker (NEC 210.19[A][3]).

11. An 18-ft. wall with a door at either end needs two receptacles (NEC 210.50[A][1]).

12. **a.** A 3-ft. wall directly inside a bedroom requires a receptacle, however, if the wall continues around a corner, you need a receptacle within the first 6 ft. of the door (NEC 210.50[A][2]).

13. A kitchen countertop needs two circuits (NEC 210.52[B][1]).

14. Kitchen counter circuits require 12-2 AWG wire (NEC 210.52[B][1]).

15. **d.** The receptacles on a kitchen counter should be 4 ft. apart (NEC 210.52[C][1]).

16. **a.** A garbage disposal requires a cord (NEC 422.16[SA][1]).

17. The smallest counter that needs a receptacle in the wall behind it is 12 in. or smaller (NEC 210.52[C][1]).

18. The GFCI receptacle in a bathroom is a 20-amp circuit (NEC 210.52[C][3]).

19. **a.** A window candle switch outlet in a side bathroom requires GFCI protection (NEC 210.8[A]).

20. Unfinished basements are referenced in NEC 210.52(G)(1).

21. Residential areas that require overhead lighting include hallways, stairways, garages, bathrooms, kitchens, and outdoors (NEC 210.70[A][1]), NEC 210.70[A][2][a], NEC 210.70[A][2][b], NEC 210.70[A][2][c]).

22. **a.** A refrigerator should run on its own circuit in the kitchen area (NEC 210.50[B], exception 2).

23. Receptacles are needed at the front and back of a home; AC must be within 25 ft. (NEC 210.52[E][1][2][3]) and NEC 210.63).

24. **a.** In a closet that is 3 ft. deep and 6 ft. wide and that has a shelf 20 in. wide, incandescent light is appropriate (NEC 410.2).

25. **c.** A 9-ft. countertop requires four receptacles (NEC 210.50[C][1]).

26. **b.** It is poor practice to install a duplex outlet for a refrigerator (NEC 210.20[D][1]).

27. **b.** Based on local codes, an exhaust fan is generally unneeded in a bathroom with a window.

28. **a.** A kitchen counter circuit is usable for a gas stove (NEC 210.50[B][2], exception 2).

29. A clothes dryer needs 10 AWG wire; the plug must be 3 ft. from the dryer (NEC 220.54).

30. **a.** A 20-amp circuit powering a washer can have more than one duplex receptacle (NEC410.10[D]).

▶ Chapter 3 Review

1. Thick-walled steel conduit is rated for a service over a roofline (NEC 230.24), although local codes also apply.

2. **b.** PVC pipe is inappropriate for overhead service, because it fails to support overhead wire (NEC 230.24).

3. c. When installing a subpanel in a 200-amp service, 100 amperes are allowable (NEC 550).
4. An electric dryer needs a neutral wire to run 120 timers and the tumbler motor (NEC 250 114.3).
5. a. For a residence, the smallest service is 100 amps, although local codes apply.
6. d. A typical residence carries 240V of electrical service.
7. In a kitchen, the garbage disposal, microwave (when hardwired and not served by a countertop receptacle), refrigerator (when the receptacle is not over a countertop), and stove receptacles require no GFCI protection (NEC 210.52[C]).
8. b. Light switches have no limit (NEC 210.70).
9. Relative to 120V service, 240V offer high voltage, low amperage, and lower power consumption.
10. According to best practice, garage receptacles should be 45 in. inches from the floor, and no less than 18 in., due to gasoline vapors.
11. a. In a bedroom circuit, 12-2 Romex is appropriate (NEC 210.2[B]). NEC does not require 12 AWG wiring in bedrooms. It is allowed when "above code" installations are desired.
12. b. Electrical service is not required in a detached garage (NEC 210.52[G]).
13. Bubble covers on outside receptacles provide weather protection when using cords (NEC 210.52[E]).
14. Split receptacles are appropriate for living rooms, family rooms, and bedrooms (NEC 210.70[A][1][2][3]).
15. From a doorway, the first receptacle should be no farther than 6 ft. (NEC 210.52[A][1]).
16. For 10-2 wire, 10-gauge ground is needed (NEC 310.4[B]).
17. b. The cubic-inch volume of a device box appears in the back middle of a plastic box and in NEC Table 314.16 for a metal box.
18. Wires should be stapled $4\frac{1}{2}$ ft. apart (NEC 314.17).
19. b. A first wire should be stapled 12 in. from a box (NEC 314.17[C]).
20. b. According to NEC, a breaker and a panel box must be specifically listed as compatible (NEC 408.53).
21. The bare wire arising from the power company side is considered grounded (NEC 230.22).
22. a. Most local codes allow electricians to use wire ties to strap wire to a three-gang box.
23. c. Per best practice, not NEC, a three-way switch accepts three wires.
24. Wire mold typically protects wires on the surface of a finished house (NEC 376).
25. In general, 80 percent of a breaker's rating should be used (NEC 220, Annex D).
26. a. All metal boxes require bonding with a ground screw and clip (NEC 215.11[E).
27. d. On a dedicated circuit using a 20-amp breaker and 12 AWG wire, a 15-amp receptacle is appropriate because it has a 20-amp feed-through (NEC 210.21).
28. In a residence, dedicated circuits include the furnace, sump pump, water heater, dryer, clothes washer, spa, microwave, and refrigerator (NEC 210.2).
29. In a typical installation, 360 degrees are allowed before a pull box is needed (NEC 362.26).
30. For the following sizes of nonmetallic sheathed cable, the following ground wires sizes are

appropriate: 14 AWG: 14, 12 AWG: 12, 10 AWG: 10, 8 AWG: 10 (NEC 310.4[B]).

▶ Chapter 4 Review

1. **b.** Given the challenges inherent in fuses, circuit breakers are easier to work with overall.
2. An ARC fault breaker is designed to detect arcing characteristics. In a residence, is it required in all areas that are not GFCI protected and, optionally, those that are not (NEC 210.12).
3. **b.** Newer-model, S-type fuses offer different threads for different fuse ratings so mismatches cannot occur upon fuse replacement.
4. GFCI and AFCI breakers have extra terminals to connect to both circuit neutral and panel neutral, whereas standard breakers connect only to the black circuit hot side directly.
5. **a.** In addition to having an identical appearance, GFCI and AFCI circuit breakers wire the same.
6. The residential areas that require GFCI protection include bathrooms (NEC 210.52[D]), garages (NEC 210.52[G][1][2]), basements (NEC 210.52[G][1][2]), storage areas (NEC 210.52[G]), and outside receptacles (NEC 210.8).
7. **a.** Garage-door receptacles require GFIC protection (NEC 210.52G).
8. A wet bar is an area with a sink (NEC 210.52[C]).
9. **d.** Per BOCA code, a home with four bedrooms on its second floor with one hall area, one bedroom on its first floor with two hall areas, and a basement, requires nine smoke detectors.
10. **b.** BOCA code prevents installation of smoke detectors on separate circuits.
11. Building Officials and Code Administrators International, Inc. (BOCA) and NFPA, not NEC, sets smoke-detector requirements.
12. **c.** The National Fire Prevention Association (NFPA) is the umbrella organization of NEC.
13. **a.** A hallway connecting a bathroom and a bedroom requires arc fault protection (NEC 210.12[H]).
14. **d.** A person can receive a lethal amount of current in as little as 10 mA.
15. A non-GFCI-protected wire from a house to a post light must be buried 18 in. deep (NEC 340.10 and references to 300).
16. **a.** GFCI protection is required for water fountains (NEC 422.52).
17. According to BOCA, 14-3 wire is required between smoke alarms because red wire signals other smoke alarms to sound, black is hot, and white is neutral.
18. **b.** Placing an electric baseboard heater below a receptacle outlet is inadvisable (NEC 424.9 FPN). Consult the manufacturer's data.
19. The fuse is the original protective device in household circuitry, although it has largely been replaced by the circuit breaker.
20. **a.** All moisture-prone outside fixtures should be marked for wet locations (NEC 210.05B).
21. To ensure safety when wiring in a hot panel, electricians should do things like wear safety glasses, don leather gloves, and use tools rated for 1000V (*OSHA Handbook, Personal Protective Equipment*).
22. **a.** BOCA codes mandate GFCI protection for bathroom candle receptacles. NEC requires protection when such receptacles are 120V (NEC 210.11[C][3]).

23. **a.** The resistance between a fuse's two terminals should be 0 ohms.

24. **a.** A hot-tub motor requires GFCI protection (NEC 680).

25. A fusible link is a thin piece of metal that has a low melting point.

26. **a.** The motor in a whirlpool tub requires GFCI protection (NEC 680).

27. Dry niche lights require no surrounding water, while wet niche lights require surrounding water to provide bulb cooling.

28. **b.** GFCI protection is not required for bathroom fan lights (NEC 210.8).

29. **a.** A bedroom fan/light requires arc fault protection (NEC 2008 210.12[B]).

30. A GFCI trips in 6 milliseconds (GFCI packing literature; see UL 943 UL Listings Book).

▶ Chapter 5 Review

1. **c.** The NEC section that describes how to size wires for conduit, as well as the conduit color-coding schema, is NEC 310 Annex C.

2. **a.** NEC Article 366.22 differentiates sheet-metal and nonmetallic auxiliary gutters.

3. NEC Sections 314.16(B)(1) and 314.16(B)(5) discuss box sizing in cubic inches, as well as the sizes needed for varied 12 AWG and communication wires.

4. **b.** NEC Table 314.16(A) NEC helps facilitate metal box sizing and cubic-inch determination.

5. **a.** NEC Sections 314.16(B)(1) to (B)(5) provide guidance when deciding when to count a grounding conductor and a neutral conductor plus device sizing.

6. NEC Section 314 provides guidance for conduit sizing, number, and location as does Chapter 9, Table 1.

7. Electricians installing wires into metal raceways and pull, junction, and panel boxes should consult NEC Section(s) 314.17(A) to (C) for insulator information.

8. **a.** NEC 314.21 provides guidance for those repairing drywall and plaster and determining the precautions governing the spacing of boxes to drywall.

9. **d.** An electrician who must make a box legal with extension rings because the box sits too far in a wall or is overfilled with wires should refer to NEC 314.22(A) and (B).

10. Varied ways to support boxes with straps, boxes, and fasteners are covered in NEC 314.23(A) to (H).

11. NEC 230.2(A) through (E) differentiate single- and three-phase services, including where and how those services must be installed.

12. **a.** NEC Sections 230.24(A) through (D) identify height requirements for service drops, but check local codes as well.

13. NEC 230.26, 230.27, 230.28, and 230.29 cover point-of-attachment requirements, but local codes also apply.

14. **d.** NEC 2008 no longer references this area.

15. **b.** To learn how to support and protect service cables, an electrician should reference NEC Sections 230.50 and 230.51(A) through (C).

16. NEC 230.54(A) through (G) are sections on the overhead services of commercial buildings and the different ways of mounting and wiring those services.

17. The reasons for enclosing or guarding service equipment appear in NEC 230.62(A) and (B).

18. NEC 230.70(A) (1–3) (B) and (C) describe disconnects sizes and how those disconnects are treated for installation.

19. **d.** Any 15-amp, AC-only switch or AC/DC 20-amp switch can be used to control a 120V motor drawing 10 amperes.

20. **b.** The proper way to ground computer and data-processing equipment is to use an isolated ground with an isolated ground receptacle (NEC 517.11).

21. **a.** Per NEC Article 250.146, unless a receptacle is an isolated ground receptacle its metal yoke is common to the screw terminal for ground.

22. For cinder fill, rigid nonmetallic conduit (RNC) is appropriate and most commonly used.

23. **b.** NEC Table 5 provides information on the volumes of different conductor types.

24. RNC should be used under sidewalks to meet installation requirements and ease handling and use.

25. Neutral current = square root of ($20^2 + 40^2 + 60^2 - [20 \times 40] - [40 \times 60] - [20 \times 60]$) = 35 amperes.

26. **a.** Per NEC Article 100, an underground location is considered a wet environment.

27. A surface raceway may extend through a partition when that raceway is a single piece.

28. **a.** As a matter of design, both elements of a hot-water heater are used all the time to provide even temperature throughout contents.

29. Because switchboards, in general, are designed for higher currents, they are physically larger than panelboards. In addition, per NEC Article 100, panelboards and switchboards have different access requirements.

30. As it pertains to switchboards and panelboards, the phrase *condition one* means no live parts, devices, or terminations are exposed (NEC Article 110.26 A1).

▶ Chapter 6 Review

1. a. 90.2
b. 220.61
c. 430.109
d. Article 358
e. 215.1
f. 230.95
g. 430.7
h. 110.26
i. 422.31
j. 356.10, 356.12, 356.20
k. Article 320.10
l. 250.119
m. 110.26
n. 408.3
o. 410.23
p. 250.20
q. 406.4
r. Article 250
s. Article 220
t. Article 330.10
u. 250.146(D)
v. 422.11
w. 240.4
x. 110.26

2. **b.** A right angle is sometimes called a square angle (NEC 314.15) and is 90°.

3. **b.** In addition to applying to Alaska and Hawaii, both outside the continental United States, per Section 90.1(D) NEC relates to international standards.

4. **d.** A reamer smoothes the inside edge of a piece of conduit after it has been cut, thereby safeguarding conductors from fraying, nicking, or other damage leading to potential hazards (NEC 362.2).

5. **c.** In addition to support every 10 ft., RMC must be supported within 3 ft. of any box or panel (NEC 344.2).

6. **c.** Vibration of mechanical equipment implies a certain flexibility that is required when supplying power to vibrating equipment (NEC 430).

7. **b.** While all three options are available as surface mount, square boxes often find use in recessed areas or installations because they have more volume (NEC 314.29).

8. **d.** Weatherproof conduit is designed to protect its contents against anything the weather might provide. RMC may be used where damage could occur, but EMT may not. Generally, metal conduit is not to be used in corrosive areas, and waterproof conduit keeps water out longer than watertight conduit (NEC 344.10[D]).

9. **a.** Plastic boxes typically lack expansion capacity because they are less rugged than their metal counterparts, but "mud ring" can be used for this purpose. Most metal boxes, whether round, square, or octagonal, can be outfitted with extension rings to increase overall box volume and thereby the number of conductors and devices the box can hold (NEC 314).

10. **b.** One master switch is not required to disconnect an entire installation, and the decision to disconnect some or all electrical loads does not necessarily come from the owner, landlord, or tenant. It is ill advised to restrict active disconnection to someone who holds a key, as doing so could prove disastrous in the event of an emergency (NEC 230).

11. **b.** A conduit bender can make offsets.

12. **d.** When no local codes specify the amount of light required for a type of occupancy, NEC must be consulted (NEC 410).

13. **b.** While fluorescent lights themselves do not dissipate heat, their associated ballasts do.

14. **a.** Gas vapors are of concern in Class I hazardous areas, but not in Class II or III. Class II and III areas are designated as such due to flammable or incendiary particulate matter that may be airborne or created as a by-product (NEC 500).

15. **d.** A grounding fitting is needed to connect a grounding conductor to a box (NEC 314.40[D]).

16. **b.** Per NEC 300.11(C), using one wireway to support another is disallowed.

17. **c.** A trapeze mount is a mount that, like a trapeze, hangs from the ceiling. Usually, a threaded rod holds the mount a fixed distance below the ceiling.

18. **c.** Single hole clamps should not be used when mounting conduit to ceilings because they only secure the conduit from one side. Two-sided clamps are preferable because they support the conduit from both sides.

19. **c.** A conduit body allows for the pulling of wires through conduit but little else. While conduit bodies are available in T and L styles, as well as straight through, little can be done with them other than pull wires. Most electricians opt to use junction boxes rather than conduit bodies simply because a junction box can be retrofitted with devices or expanded to allow for more conductors.

20. **b.** All unused openings should be sealed (NEC 408.7). When ventilation or dampness is an issue, other solutions, like ventilation fans or rated equipment, are preferred.

21. **d.** Bending conduit creates a shorter path between points and uses less conduit. The run is easier to construct because no L fitting is required.

22. **a.** The limit is 13 receptacles on a commercial branch circuit (NEC 220).

23. **d.** For exit signage, the required power allocation is 1,200W (NEC 220.14[F]).

24. **b.** Schedule 80 PVC, being more robust than its Schedule 40 counterpart, has thicker walls. Thicker walls mean smaller inside volume but greater protection against physical damage.

25. **a.** Most air-conditioning systems have three motors. The compressor motor is three phase, while the condenser and fan motors are single phase (NEC 440).

26. Per NEC article 80.19(F) of the 2005 National Electrical Code, when any electrical installation under a permit is completed, the electrical contractor must notify the electrical inspector. This reference was removed for the 2008 code.

27. When wiring a hot panel, an electrician should wear safety glasses, don leather gloves, and use tools rated for 1,000V (*OSHA Handbook*).

28. **a.** Per NEC 210.11, loading per unit area must be evenly distributed among branch circuits.

29. **c.** In Class II locations, *combustible dust* is present in sufficient quantities to produce a hazard of explosion or ignition. Division 2 means this hazard occurs under abnormal conditions. In a wood shop, for example, this condition may exist if the wood-shavings vacuum system becomes clogged, resulting in a heavier-than-usual concentration of wood particulates in the air (NEC 500.5).

30. **b.** NEC Table 402.5 defines the ampacity requirements for fixture wires. Note that the gauge of many fixture wires is smaller than wires and conductors used in branch circuitry.

▶ Chapter 7 Review

1. Standard breakers and GFCI breakers differ in that GFCI breakers are marked as such.

2. Raceways must be installed $1\frac{1}{2}$ in. below decking, although this does not apply to RMC (NEC Section 300.4).

3. **d.** When referencing NEC, it is important to know that 750 is not a standard amperage value (NEC 240.6).

4. **b.** Panel boards are not limited to 42 protective devices (NEC 408.54).

5. **a.** An electric water fountain in a public building requires GFCI protection (NEC Section 422.52).

6. **d.** Electric pipe or plumbing heaters (wraparounds) require signage adjacent to supplied or supplying equipment, as well as every 20 ft. (NEC Section 427.13).

7. **a.** Overcurrent protection schemes may use relays (NEC 430.32).

8. **b.** There must be a means of disconnecting motors that is not subject to automatic closure (NEC 430.100). For instance, it is illegal to hard-wire a sump pump so that the only way to inactivate it is to hold the plunger in the down position. There would have to be a separate switch to disconnect the entire apparatus.

9. **d.** Motors operated at greater than 600 volts require lockout-tagout systems (NEC 430.227).

10. Greenfield is flexible metal conduit (FMC) (NEC Article 360.2).

11. **a.** To be prepared for commercial installations, electricians should carry portable lockout-tagout sets.

12. **a.** NEC no longer provides guidelines for connecting to an electrical supply. Check with your local authorities for specific guidelines.

13. **b.** In a continuous load, rated current is expected to flow for at least 3 hours (NEC Article 100).

14. The phases in a three-phase panelboard must be arranged in a particular order, such as A, B, C from left to right or top to bottom as seen from the front (NEC 408.3).

15. **c.** NEC Chapters 1–4, which govern basic practices, are in effect in hazardous areas, except as modified by specific requirements outlined in NEC Chapter 5.

16. **a.** Vending machines need GFCI protection (NEC 422.51).

17. **b.** NEC Article 314.4 requires all metal boxes to be grounded.

18. Per NEC 408.4, a panelboard's circuit directory must legible so other electricians or emergency personnel can decipher it.

19. **b.** Protecting a motor and protecting a motor circuit are two very different things. NEC Article 430 Part III outlines the requirements for motor protection. NEC Table 430.37 is an excellent reference.

20. **b.** All unused panelboard openings should be sealed (NEC 408.7). When ventilation or dampness is an issue, other steps, such as installing ventilation fans or using rated equipment, should be taken.

21. NEC Table 300.5 provides the requirements for buried cable of up to 600 volts.

22. **b.** Appliances with motors over 0.125 HP must have a separate means of disconnect (NEC 422.31).

23. **c.** Although NEC 450.46 requires only a drain in the event of transformer action exceeding 100KW, it is good practice to be able to contain and clean a spill of dielectric.

24. **a.** According to the NEC, equipment must be installed according to manufacturer's instructions and specifications. When it is not, an apparatus's listing is void and the installation violates NEC.

25. To calculate a commercial branch circuit that could effectively handle 20KW, per NEC Table 220.42 and 220.44, use 100% of the first 10KW and 50% of the demand above 10KW. In this case, the demand factor would be 10,000 plus half the remaining 10,000 or 15,000W.

26. **b.** Direct-current motors are not always protected at a minimum of 150% of full load current because (1) the percentage represents a maximum, not a minimum; and (2) if the protection is an instantaneous breaker, the percentage used is 250% (NEC Table 430.52).

27. Per NEC Article 100, the phrase *in sight* means visible and within 50 ft.

28. **b.** Circuit breakers rated at 100 amperes or fewer and 600 volts or fewer must have their ampere ratings molded, stamped, etched, or similarly marked into their handles or escutcheon areas (NEC 240.83[B]).

29. **b.** Receptacles as specified in NEC Article 647.7(A)(2) may not be used for lighting or motors.

30. **b.** Per 376.56(A), wireway splices and taps in a wireway must be accessible to ensure safety.

▶ Chapter 8 Review

1. In a typical residence, three items that typically use PLC technology are HVAC systems, security systems, and lighting and audio-control systems.

2. **a.** One may argue that relays are current controlled (they require a minimum current, called holding current, to stay energized), but current cannot exist without voltage.

3. **d.** While primary jurisdiction over industrial electrical installations resides with local agencies, those agencies refer to other agencies, like NEC and NFPA, and their publications (NEC Article 80).

4. **a.** Digital systems rely on 1s and 0s to make decisions and to control hardware based on those decisions.

5. **a.** Sensors like thermometers or thermocouples measure static or unmoving quantities.

6. Processing time is the time it takes for a PLC to take action once the appropriate signal is received.

7. **d.** The generic term for a device that receives signals, makes decisions, and generates control signals based on its inputs is *computer*. Each computer has a central processing unit (CPU), which is central to the PLC. A PLC and computer serve similar purposes.

8. **d.** To calculate power factor when a system's total power is 10KVA, 6KVA resistive, divide 6KVA by 10KVA and multiply by 100% for 60%.

9. **c.** Because the system in Q7 has a low power factor due to excess capacitance, the solution is to add an inductor because inductance counteracts capacitance.

10. **b.** Most industrial entities like plants and factories employ in-house electricians to work with local authorities to plan, execute, and maintain any electrical installations, conversions, or upgrades.

11. **b.** Typically, peak demand occurs in the early morning when employees arrive at work to start equipment, which requires the most current.

12. Scan time is the time a PLC needs to complete an iteration of its program.

13. **a.** To be programmable over the Internet or through a network, a PLC requires an IP address, the protocol by which hardware is networked.

14. **a.** A normally open switch has an inner spring to keep the switch deactivated.

15. A normally closed switch sits on a shelf with its contacts connected. Both normally open and normally closed switches have springs to keep them in one position when not being pressed.

16. **b.** A tachometer, which measures the speed of a moving object, is a transducer.

17. A variable-speed drive is a type of motor controller used in newer HVAC systems to boost efficiency.

18. **a.** The steps hardware takes constitute a program.

19. **c.** The time it takes to physically move a switch when given a command is called control lag.

20. Discrete systems rely on individual components, whereas integrated systems rely on combined ones.

21. c. Kinetic energy, or energy in motion, and is related electrically to current.

22. PLCs and security systems communicate using voltages between 5 and 24.

23. a. In typical electrical circuits, natural frequency is 60Hz.

24. A relay is a device that uses low power signals to control large electrical loads.

25. d. Step-down transformers increase available current.

26. a. A limit switch senses the presence or absence of a part contacting its surface limit. When the object reaches the switch, something happens. In a garage door opener, for example, when the door itself hits a switch at the top of its travel, the motor turns off.

27. Snapshot, or scene, is the current status of a system.

28. a. In most digital systems, YES = 1 and NO = 0.

29. A static system or parameter does not change, while a dynamic system or parameter does.

30. b. The NEC maintains jurisdiction over most industrial installations, but in many cases local guidelines which are "tighter" will supercede the NEC's minimum acceptable standard.

▶ Posttest 1 Review

1. c. Per NEC 240.4(D)(1), 18 AWG copper conductor protection cannot exceed 7 amperes. Changes in 2008 code include the use of 16 and 18AWG copper.

2. Equipotential bonding is installed primarily to reduce voltage gradients in a pool area (NEC 680.26[A]).

3. b. Batteries, flywheels, and superconducting magnetic storage equipment are not part of hybrid electrical systems (NEC 705.2).

4. NEC Article 708.2 defines the terms represented by the acronyms SCADA, COPS, and DCOA.

5. b. Up to 18 in. above the floor, an aircraft hangar, and its associated areas, is considered a Class1, Division 2 or Zone 2 location (NEC513.3[B]).

6. In addition to 12-2 wire, a dining room circuit must use 20-amp service (NEC 210.11C, 210.24, 240.4).

7. a. When less than 24 in. deep, the underground wiring for a warehouse must be routed in metal conduit (NEC 310).

8. b. An exhaust fan is unneeded in a bathroom with a window (local codes prevail).

9. d. Per NEC 90.2(A) and (B), only fiber optic cable falls under NEC jurisdiction.

10. a. In certain cases, such as when code revisions call for new materials that are not yet available (e.g., colored conduit), AHJs may waive parts of NEC (NEC 90.4).

11. NMC cable has a flame retardant and moisture-, fungus-, and corrosion-resistant nonmetallic covering (NEC 334.02).

12. b. The demand factor in NEC 220.61(B) is not allowed when the demand factors in NEC Table 555.12 apply (NEC 555.12)

13. NEC 408.4 requires circuit directories for all panel boards.

14. c. PLFAs cannot be sourced by GFCI protected receptacles (NEC 760.121[B]).

15. b. Electricians should consider all devices active, or live, until they determine otherwise.

16. Metallic raceway requirements appear in NEC 386; requirements for nonmetallic raceway appear in NEC 388.

17. a. At most, a kitchen disposer cord can be 36 in. (NEC 422.16[B][2]).

18. An outlet gives power to other power-drawing devices (NEC 100).

19. a. NEC Table 310.16 provides information on the minimum allowable ampacity of THHR.

20. a. As long as the service does not exceed the roof, PVC is acceptable per local codes in most jurisdictions.

21. b. Although central heating equipment is considered an appliance (NEC Article 100), it must be on its own branch circuit (NEC 422.12).

22. c. The controller can be an attachment plug and receptacle for a portable motor rated at $\frac{1}{3}$ horsepower and lower (NEC 430.81[B]).

23. At 1.5 amps per receptacle, 12 receptacles can resides on a 20-amp circuit (NEC 210).

24. a. In general, Class 2 locations should have no explosion-proof equipment or wiring (NEC 502.5).

25. d. The ampacity of capacitor circuit conductors should not be less than 135 percent of a capacitor's rated current (NEC 460.8[A]).

26. This symbol represents a countertop receptacle.

27. a. (NEC 404.14).

28. b. Per revisions to NEC 250.36, a high-impedance grounded system can be installed for reasons other than a need for continuous service.

29. a. Branch-circuit overcurrent device ratings should align with the protective ratings of appliances (NEC 422.11[A]).

30. b. The insulation color of parallel conductors need not match (NEC 310.4[B]).

31. b. In air-conditioning and refrigerating equipment, the marked rated-load current exceeds the branch-circuit selection current (NEC 440.2).

32. In bus bars and conductors for motor circuits and controllers, the B phase has the higher voltage to ground on 3-phase, 4-wire, delta-connected systems (NEC 430.97[B]).

33. a. In show windows, chain-supported luminaires can be externally wired (NEC 410.14).

34. NEC 356.30 is the reference when lengths of liquid-tight nonmetallic conduit exceed 6 ft.

35. a. Garbage disposals require cords (NEC 550 and 551).

36. The feeder is all circuit conductors between the service equipment, the source of a separately derived system, or other power supply source and the final branch-circuit overcurrent device (NEC Article 100).

37. c. The induction type of pipeline heater uses hysteresis (NEC 427.2).

38. b. The requirements in NEC 314.27(A) and (B), which cover ceiling fans with luminaires, are less stringent than those in NEC 314.27(D), which covers ceiling fans with no luminaires. Because ceiling fans move and wear their mounts, they cannot be supported solely by boxes unless those boxes are so listed.

39. Two, one in front and one in back (NEC 210.52[E][1][2][3] and 210.63).

40. b. The proper way to ground computer and data-processing equipment is to use an isolated ground with an isolated ground receptacle (NEC 517).

41. The requirements for permanently connected appliances appear in NEC 422.31.

42. **d.** Thermoset's maximum operating temperature is 90°F (NEC Table 310.13[A]).

43. **b.** Cables that are rated properly may be installed using directional boring (NEC 300.5[J]).

44. **b.** Nonmetallic cable branch circuits are allowable in structures lacking height limitations according to NEC 590.4(C).

45. **b.** Rather than a best practices handbook, the NEC is a set of requirements for electrical work (NEC 90.1[C]).

46. NEC Table 1, Chapter 9, sets out the fill requirements for ENT (NEC 362.22).

47. **d.** NEC 100 provides the definition for the phrase *in sight*.

48. **b.** Illumination can be manually controlled in electrical equipment rooms (NEC 110.26[D]).

49. Using NEC Chapter 9, Table 4, determine the total area of the existing raceway. Then, multiply that figure by 40% to calculate the wire fill allowed for the raceway. Using NEC Chapter 9, Table 5, look up the wire type you plan to use, and total the square inches for all planned conductors. Ensure the result is less than 40% of the raceway's area.

50. The load in an electric circuit converts electrical energy to heat, magnetism, or light.

51. **b.** A forming shell is a structure that mounts a wet-niche luminaire in a pool or spa (NEC 680.2).

52. **b.** A sump pump, like a refrigerator, should continue running when other devices fail but in some locations it is true because of the local code requirements.

53. **b.** In terms of electrical work, nursing home sleeping rooms do not qualify as patient-care areas (NEC 517.10[B][2]).

54. **b.** NEC addresses explosion properties and ignition temperature separately (NEC 500.8).

55. **b.** Because a receptacle or switch counts as two conductors (NEC 314.6[A]) and 14 ga. wire conductors each require 2 cubic inches, a receptacle or switch receiving 14-2 wire counts as four cubic inches (2 conductors × 2 cubic inches).

56. **a.** As temperature increases, PVC grows more than RTRC (NEC Tables 352.44 and 355.44).

57. The bare wire arising from the power-company side of electrical service is considered neutral.

58. **c.** Knob and tube wiring is appropriate when repairing or enhancing existing installations (NEC Article 394).

59. **a.** Electricians can install feeders with a common neutral (NEC 215.4).

60. Splices in knob-and-tube wiring should be soldered (NEC 394.56).

61. **a.** All bathroom receptacles require GFI protection, even in side bathrooms (NEC 210.8).

62. NEC 250.134 covers permanent wiring procedures for ground equipment, while NEC 250.138 provide cord and plug wiring requirements.

63. **a.** When more than one receptacle or load is on the branch circuit for a recreational vehicle, a 15-ampere receptacle can be protected by a 20-ampere fuse or circuit breaker (NEC 551.43[C]).

64. **b.** NEC 410.130(e) states the thermal protection requirements for fluorescent lamp information.

65. **b.** Conduit cannot be stapled.

66. **b.** Unused switchboard openings must be covered to provide as much protection as that on the wall of the enclosure (NEC 408.7).

67. **b.** Minimum sizes of equipment grounding conductors appear in NEC Table 250.122.

68. **b.** A revision to NEC 2008 allows corrosible metals like iron to be used as grounding electrodes when they are treated or anodized to resist corrosion.

69. **a.** According to NEC, expansion should be factored into a system's design specification (NEC 90.8).

70. When a new residence is wired, the builder, homeowner, inspector, and job manager (boss) should all be satisfied with the work.

71. **b.** Range hoods, which need not be hardwired, can be used as receptacles (NEC 422.16[4]).

72. **b.** The top of a switchboard must be 3 ft. from the ceiling (NEC 418.18A).

73. **b.** Although NEC 450.1 states that it covers all transformers, eight exceptions apply.

74. **c.** Hydrogen is a Class 1, Group B, material (NEC 500.6[A][[2] fpn).

75. **b.** Per NEC 230.33, which details the splicing requirements for service laterals, service lateral conductors may be tapped or spliced.

76. **a.** According to NEC 210.8 and 590.(6)(A), feeders supplying 15- and 20-amp receptacle branch circuits can be protected by ground-fault circuit interrupters.

77. In Table 250.3, Sections 553.8, 10, and 11, NEC discusses the grounding requirements of floating buildings.

78. **a.** Given that carnival equipment is outside, it requires GFCI protection (NEC 525.23[A]).

79. **c.** NEC Article 430 helps electricians calculate continuous and noncontinuous loads.

80. **d.** When seven elevators are installed on one feeder, the demand factor is 0.77 (NEC 620.14).

81. **b.** To measure current, a meter must be in a circuit (i.e., in series with the load through which it is measuring current).

82. According to GFI Packing literature, a GFCI will trip in 6 milliseconds (see UL 943 of the *UL Listings Book*).

83. **c.** In a continuous load, rated current is expected to flow for at least three hours (NEC Article 100).

84. **b.** Per fire codes, smoke detectors cannot reside on separate circuits.

85. **a.** Attics with furnaces require receptacles (NEC 210.63).

86. Among the information included in conductor markings is maximum rated voltage, conductor type, manufacturer, and conductor size by AWG.

87. **c.** A ground ring is buried a minimum of 30 in. below the ground (NEC 250.33[2][F]).

88. **b.** While NEC has no height requirements for dryer plug receptacles, the suggested height is 45 in. (NEC 422).

89. This symbol represents a weatherproof outlet, which is usually found outside the home.

90. **a.** Residential wiring schemes may use floor receptacles (NEC 210.52[A][3]).

91. The principle of power distribution distinguishes between the disconnect, or main panel, and the subpanel, which is found inside a residence (NEC 220).

92. **a.** The S_3 notation indicates a three-way switch.

93. A 200-amp breaker must be fed by 4/0 aluminum or 3/0 copper wire (NEC Table 250.122).

94. **d.** The circuit for a GFCI bathroom receptacle is 20 amps (NEC 210.52[C][3]).

95. **b.** The exceptions to this requirement appear in NEC 230.94

96. **b.** Aluminum conductors cannot terminate within 18 in. of the earth (NEC 250.64[A]).

97. NEC 220 Part IV covers the computation of farm loads and is referenced most often when calculating loads for farm applications.

98. An 18 ft. wall with a door at either end requires two receptacles (NEC 404.15).

99. **b.** NEC 250.4(B) covers ungrounded systems.

100. **b.** In today's world of do-it-yourselfers and old homes built to old codes, junction boxes may reside behind drywall. Contemporary electricians should expect the unexpected, which means "call before digging" and "check before cutting."

101. NEC 110.14 contains the requirements for electrical connectors.

102. **b.** If for every 8 volts into the transformer 1 volt out comes out, the output is $\frac{1}{8}$ the input (120/8) or 15 volts.

103. **d.** The integrity of the conductor must be maintained throughout its travel, therefore, using insulation of a similar rating insures this integrity (NEC 110.14[B]).

104. Conductor ampacity determines a circuit's maximum ampacity.

105. **a.** A surface raceway may extend through a partition when the raceway is a single piece (NEC 394).

106. **c.** Four wires are needed from a main panel to a subpanel in the same building: 2 hots, 1 neutral, and 1 ground (NEC Article 110).

107. **b.** No receptacle is needed when there are 2 in. of countertop behind a sink (NEC 210.50).

108. **a.** GFI protection is needed when an exhaust fan and light are installed in a shower area (NEC 411.7).

109. NM, NMC, and NMS cables may be used in residential applications (NEC 334.10).

110. **d.** NEC 386.56 and 388.56 state that electricians may not exceed 75% of box fill when splicing and tapping box conductors.

111. NEC 210.52(A) details these requirements.

112. **b.** While NEC 250.8 lists seven ways to connect grounding and bonding equipment, an eighth entry states "other listed means."

113. **b.** NEC Table 220.103 is used to determine farm load demand.

114. **c.** Class II locations have enough combustible dust to produce explosion or ignition hazards; Division II means these hazards occur under abnormal conditions (NEC 500.5).

115. When a portable optional standby source is used as a separately derived system, it shall be grounded to a grounding electrode (NEC 702.10[A]).

116. NEC 210.8 identifies eight locations requiring GFCI protection.

117. Per NEC 210.11, loading per unit area must be evenly distributed among branch circuits.

118. **a.** NEC 250.146(D) provides requirements for running isolated computer receptacles.

119. **b.** Although NEC 210.52 requires receptacles for all balconies and decks that are accessible from inside dwellings, it exempts balconies and decks under 20 sq. ft.

120. **a.** All outside fixtures that get wet or damp should be marked for wet locations.

121. **c.** NEC Article 424 sets out the requirements for fixed electrical space heating equipment.

122. According to local codes, underground meter bases must be 5 ft. high from finish grade.

123. **a.** All protection devices must be accounted for in the panelboard.

124. NEC Sections 312.6, 312.7, and 312.8 discuss cabinet clearances for bending and splicing of conductors.

125. Split receptacles can be installed in living rooms, family rooms, and bedrooms (NEC 210.70[A][1][2][3]).

126. **c.** NEC 110.26 2 dictates that all working spaces with electrical equipment have lights.

127. **a.** NEC Article 220 identifies minimum lighting loads for three categories of occupancy.

128. **c.** NEC Article 770 contains conduit fill requirements for fiber optic cabling.

129. In a residence, areas like the kitchen (NEC 210.52[A][1]), bathroom (NEC 210.11[C][1][2]), and laundry room (NEC 210.11[C][1][2]) require 20-amp circuits.

130. **b.** Pendant-supported enclosures must comply with NEC 314.23(H).

131. **a.** NEC 220.3, B9, provides information for electricians calculating loads for receptacles on the trap.

132. **d.** Switchgear and control-assembly frames shall be connected to a(n) equipment grounding conductor or, where permitted, the grounded conductor (NEC 490.36).

133. **a.** Lights must illuminate crawl spaces if used for storage (NEC 2008 210.70[A][1]).

134. An electric dryer needs a neutral to run 120V timers and its motor.

135. NEC addresses transformer installations into two voltage levels in NEC Article 450: Table 450.3(A) for over 500 volts and Table 450.3(B) for under 500 volts.

136. **a.** When wires extend from residences to post lights without GFI protection, they must be buried 18 in. deep (NEC 300.9).

137. **b.** Liquid-filled transformers have differing requirements based on the dielectric fluid they use (NEC 450.23, 450.24, 450.25, and 450.26).

138. **d.** Deicing systems are discussed in NEC Article 426, not NEC Article 526.

139. **c.** Variable-speed motors require protection for overtemperature and overcurrent (NEC 430.126).

140. **a.** NEC 230.71(A) states that up to six disconnects may be used to turn off power to a building.

141. **a.** There must be two or more receptacles on the circuit.

142. A 200-watt bulb in a 300-watt lighting fixture still calculates as 300 watts (NEC 220.3[B][4]).

143. **b.** Motor overload protection is based on the motor nameplate rating in compliance with NEC 430.32.

144. **b.** A surge arrester over 1 kV can be installed when the rating of the surge arrester is less than the maximum continuous phase-to-ground power frequency voltage available at the point of application (NEC 280.2).

145. Provisions for sensitive electronic equipment as discussed in NEC Article 647.3(1) apply only to industrial installations.

146. **b.** Per NEC 647.7(A)(2), receptacles may not be used for lighting.

147. Switchboards, in general, are designed for higher voltages (over 600V) (NEC Article 100).

148. **d.** NEC Article 240 Part 6 governs cartridge fuses.

149. **a.** When a suitable substitute is available, such as an underground pipe system, electricians may opt to use no grounding electrode (NEC 250).

150. **a.** Fire code dictates that smoke detectors reside on AFI breakers.

151. **b.** Audio signal-processing requirements are discussed in NEC Chapter 6, specifically NEC 640.23 and 640.24.

152. NEC 220.12 addresses track lighting, as well as the show windows used in shops.

153. **c.** NEC Article 725 governs the installation of remote controls and signal circuits.

154. Class I power source limitations for alternating current appear in NEC Chapter 9, Tables 11A and 11B.

155. **b.** There is no burial requirement (NEC Table 300.5).

156. **a.** NEC 725.54 provides reference material when installing low-voltage wires with a 120 volt power line.

157. This symbol represents "recess."

158. **b.** A water heater is considered a continuous load (NEC 422.13).

159. NEC Article 440 covers requirements for air-conditioning and refrigeration equipment.

160. NEC Article 440, Section 240.6, covers the installation of branch circuits over current devices with ratings not to exceed the MOP.

161. **b.** NEC Article 354 prohibits the use of NUCC in exposed locations and inside buildings.

162. NEC 250.4(A)(5) states that the Earth is not to be the only equipment grounding conductor.

163. The phrase *shall be* is synonymous with *must be* (NEC 100).

164. **d.** FEP, FEPB, MTW, PFA, THHN, THHW, THWN, and a host of other insulated conductor types are allowed in dry/damp locations (NEC 310.8[B]).

165. **a.** NEC Tables 310.67 through 310.86 compare gauges and configurations of conductors as run in Earth and in conduit.

166. NEC 250.118 covers equipment grounding conductors for copper aluminum or copper-clad aluminum and several types of conduit and metallic tubing.

167. **a.** Heart fibrillation is the major cause of death from electrical shock.

168. **b.** The smallest counter requiring a receptacle is 12 in. (NEC 210.52[C][1]).

169. NEC Article 310.8 lists which conductor insulation types are appropriate for wet locations. NEC Sections 310.9 and 310.13 provide specific information on selected types.

170. **b.** In general, Type AC cable should be secured within 12 in. of every outlet box, junction box, cabinet, or fitting and at intervals not exceeding $4\frac{1}{2}$ ft. where installed on or across framing members (NEC 320.30).

171. **b.** Red wire, or 14-3, signals other smoke alarms to sound.

172. **d.** Garbage disposal, refrigerator, and stove receptacles require no GFCI protection (NEC 210).

173. **a.** It is appropriate to use 12-2 Romex in bedroom circuits.

174. Circuit ampacity must not be less than 125% of the continuous load plus 100% of the noncontinuous load (NEC 210.19).

175. **d.** High voltage, low amperage, and low power loss are all benefits of 240V service.

176. **b.** NEC 110.23 states, "Unused current transformers associated with potentially energized circuits shall be short circuited."

177. **b.** NEC allows any parties to suggest code amendments.

178. NEC Chapter 4, Table 8, covers the resistance of copper and aluminum wire.

179. Per NEC Chapter 9, Table 8, a grounded conductor applies as a voltage ground, not a grounding path.

180. **c.** Valance lighting fixtures are placed close to ceilings.

181. **b.** Appliances with motors over $\frac{1}{8}$ hp (0.125) must have a separate means of disconnect (NEC 422.31).

182. AFCI is used to protect from fire, while GFCI is used to protect people.

183. **c.** When outdoors, overhead conductors must be protected within 10 ft. of a building (NEC 225.4).

184. NEC Article 100 defines continuous loads.

185. **b.** The Earth is not an efficient ground fault current path (NEC 250.4(A)(5) and 250.4(B)(4).

186. **c.** Open conductors on separate circuits must be separated by 4 in. (100 mm) (NEC 225.14).

187. NEC 250.2 covers ground fault current path.

188. **c.** At minimum, 3 VA should be allocated for each square foot of a residence (NEC 220). In certain areas, like the kitchen, which may require better lighting, for example, a higher allocation may be warranted.

189. **a.** Connection is a requirement for electrically conductive materials that are likely to become energized (NEC 250.4[A][4]).

190. Because resistance must be less than or equal to 25 ohms (NEC 250), most local codes require two ground rods for service. Those rods must be at least 8 ft. long and 6 ft. apart.

191. **b.** NEC 230.24 dictates that basic overhead service is 12 ft. off the ground.

192. **b.** Using one wireway to support another is disallowed (NEC 300.11[C]).

193. NEC 250.22 lifts the grounding requirement for circuits.

194. **b.** Fuses on the load side of a transformer will not protect the transformer. Per NEC 450.3, install OCPD at the transformer primary when it is enclosed.

195. **a.** The handhole enclosures used in underground systems should identified (NEC 314.30).

196. **a.** Generally, temporary service is subject to the same requirements as permanent service (NEC 527).

197. NEC 250.24(B) mandates bonding for enclosed service disconnects.

198. **a.** Voltage that is 1,000 or lower must run to each service disconnecting means and bond to each disconnecting enclosure (NEC 250.41[C][1] to [C][3]).

199. In a revision to 2008 NEC, which refers to the overhead height as the finished or completed grade rather than "ground level" as in previous editions, clearances for overhead conductors apply to finished or grade level.

200. **c.** NEC Article 424, Part 5 contains information related to cable-style fixed space heaters.

201. **b.** A building may act "on its own" as a grounding system and therefore cannot serve to augment another grounding system (NEC 250.52[A][A]).

202. **a.** NEC Article 240 covers circuit protective devices.

203. **a.** The name or acronym TVSS has been replaced with a more generic term, SPD, which represents "surge protection device."

204. Except in some cases, only conductors of size 1/0 may be installed in parallel (NEC 310.4[A]).

205. **d.** NEC Table 250.65 provides guidance when there is no outer current protection.

206. **a.** To avoid such hazards as draped extension cords, 4 ft. kitchen islands require receptacles (NEC 210.50).

207. **a.** The wire that goes on the common screws of a three-way switch is black or red with black tape.

208. **b.** For safety, a metal fence in public use must connect to a ground rod (NEC 250).

209. **a.** A nonmetallic wireway can be filled to more than 20% capacity in most cases (NEC 377).

210. **b.** Install an OCPD in series with each ungrounded conductor (NEC 240.20).

211. Per NEC 230.40, "Every service drop or lateral shall supply only one set of service-entrance conductors."

212. **b.** Switches must be installed so that loads cannot be activated by the force of gravity, which means "off" must be down (NEC 404).

213. NEC 410.67(C) outlines the requirements for running tap conductors from boxes to light fixtures.

214. **a.** When installed in concrete or masonry floors, constant wattage heating cables should not exceed $16\frac{1}{2}$ watts/linear foot of cable (NEC 424.44[A]).

215. **a.** Box volume should not exceed the fill (NEC 314.16).

216. The wire on kitchen counter circuits must be 12-2 AWG (NEC 210.52[B][1]).

217. **b.** A refrigerator circuit counts as a counter circuit (NEC 250).

218. NEC NFPA 30 outlines the requirements for installing parking lot lights around a building.

219. **b.** NEC Table 250.122 helps electricians size separate insulated isolated equipment grounding conductors.

220. **a.** Lighting circuits can reside in more than one room (NEC 220).

221. **b.** When the surface area of a kitchen island is 2 sq. ft. or more, the island requires a receptacle (NEC 210.50).

222. **a.** A feeder's computed load must exceed the computed load of the circuits it feeds (NEC 220.10).

223. A device box displays its cubic inches on its back middle surface (NEC 314).

224. A breaker should operate at 80% capacity.

225. In residences, bathrooms (NEC 210.52[D]), garages (NEC 210.52]G][1][2]), basements (NEC 210.52[G][1][2]), storage areas (NEC 210.52[G]), and outside receptacles (NEC 210.52[E][3]), NEC 210.8) require GFCI protection.

226. **d.** NEC Article 90 explains the scope and application of NEC.

227. **a.** Vending machines require GFCI protection (NEC 422.51).

228. NEC Article 430.6 discusses how to determine motor ampacity ratings, including information on correct table use.

229. **a.** Multiwire branch circuits are acceptable in Class I environments only when all ungrounded conductors are switched simultaneously to prevent transient arcing (NEC 210.4).

230. **a.** A controller may serve as the disconnect, per conditions laid out in NEC 427.55 and NEC 427.56.

231. **a.** The operator starts, stops, and controls an amusement ride or supervises a concession (NEC 525.2).

232. **c.** (NEC Table 430.248)

233. **d.** Section 450.9. When transformers must be ventilated, as when they are installed in closed areas, NEC 450.9 is the appropriate reference.

234. **b.** NEC 110.27 provides the guidelines for guarding live or energized components.

235. **a.** The "3 ft. rule" for device clearances is a minimum because, depending on the location and environment, 3 ft. of clearance may be insufficient.

236. **b.** Per NEC 408.3, the high-voltage leg must be identified as the B leg.

237. **a.** While NEC 450.46 requires a drain only when transformer action exceeds 100KW, such a drain is a good idea so as to contain and clean dielectric spills.

238. **b.** The handles or escutcheon areas of circuit breakers rated at 100 amperes or less and 600 volts or less should be molded, stampled, etched, or similarly marked with their ampere ratings (NEC 240.83[B]).

239. **b.** Instead of protecting against lightning, TVSSs are designed to protect against known and predictable anomalies, including short-duration spikes of high voltage, but only those seen as possible side effects of hardware issues.

240. Per NEC 210.52, receptacles along a long wall must be within 12 ft. of each other so that no point along the wall is more than 6 ft. from a receptacle.

241. **a.** Per NEC Table 220.13, start the calculation by using 100% of the first 10 KW and 50% of the demand above 10 KW. In this case, the demand factor would be 10,000 plus half the remaining 10,000 or 15,000 W.

242. **c.** A fire alarm system installed in EMT using metal boxes requires no AFCI protection (NEC 210.12[B]).

243. **a.** NEC Table 250.66, and specifically subscript "a," applies to the derived conductors of separately derived AC systems.

244. **a.** Receptacles should be at least 6 ft., measured horizontally, from the inside walls of a spa or hot tub (NEC 680.43[A][1]).

245. **b.** Circuit breakers must be able to be manually turned off (NEC 240.80).

246. Fixed electric space-heating loads should be calculated at 100% of the total connected load (NEC 220.51).

247. **c.** Insulated wire must connect the grounding terminal of a hospital-grade receptacle to ground (NEC 517.19[B][2]).

248. **d.** Per NEC Article 100, a "qualified person" is knowledgeable and skilled in construction and equipment operation, as well as trained in related hazards.

249. b. The motors, controllers, and wiring of electrically operated pool covers should be no less than 5 ft. from the inside wall of the pool unless they are separated from the pool by a wall, a cover, or another permanent barrier (NEC 680.27[B][1]).

250. a. NEC considers procedure tables in critical care areas patient bed locations (NEC 517.2).

▶ Posttest 2 Review

1. c. A drinking fountain is not considered a fountain by NEC terms (NEC 680.2).

2. b. Any part of the direct-buried equipment grounding conductor that runs to a livestock building must be insulated or covered copper (NEC 547.9[D]).

3. NEC Article 250 covers grounding electrode systems.

4. Switches must be marked for voltage, current, and horsepower (NEC 404.15[A]).

5. b. NFPA 70 is the same as NEC, so the two do not differ.

6. d. All these statements about effective grounding paths are true (NEC 250.68[B]).

7. a. As part of an FPN after the definition of "authority having jurisdiction (AHJ)," property owners can act as their own AHJs (NEC Article 100).

8. According to NEC Section 406.3(D)(3)(a), electricians may replace defective, ungrounded receptacles with like replacements.

9. b. Electricians may install inductive electric heaters (NEC 424 Part V).

10. a. The branch circuit conductors supplying one or more units of a data-processing system should have an ampacity not less than 125% of the total connected load (NEC 645.5).

11. b. NEC Article 314.4 requires grounding for all metal boxes, regardless of the presence or absence of a grounding conductor.

12. b. NEC 410.116 specifies that a recessed luminaire not identified for contact with insulation must be no less than $\frac{1}{2}$ in. away from combustible materials.

13. a. IMC with a trade size of 1 in. has a 27 metric designator (NEC Chapter 9, Table 4, Article 342).

14. a. Bonding with ground screws and clips is required for all metal boxes (NEC 250).

15. a. Secondary electrical distribution systems to mobile-home lots should be single-phase 120/240, nominal (NEC 550.30).

16. b. Because a finished basement is a livable space like any other room in a residence, no GFCI protection is needed (NEC 210.8).

17. Class I locations have hazardous combustible dust (NEC 500.5[B]).

18. a. Each load is connected in a separate path in a parallel circuit.

19. d. Trade size $\frac{1}{2}$ ENT has a nominal internal diameter of 0.560 in. (NEC Chapter 9, Table 4, Article 362).

20. b. While a raceway may traverse a high temperature area, its terminations may reside in an area of lower temperatures.

21. c. A 10 AWG wire holds 30 amps (NEC 310.16).

22. **b.** Electricians must take expansion contraction into account when heating elements and assemblies bridge expansion joints (NEC 426.21[C]).

23. This symbol represents a split or half-and-half receptacle.

24. NEC Article 408 outlines the requirements for construction and panel application.

25. **b.** While NEC had no mandate for the installation of DC bonding jumpers in 2005, the same is not true in NEC 2008 (NEC 250.168).

26. **b.** Upon receiving a request to add lighting to a job, an electrician should first determine how much added loading the supplying circuit can deliver.

27. NEC Sections 210.63 and 210.8(B)(2) target rooftop receptacle outlets.

28. **b.** Electricians cannot install breakers in panel boxes that differ from manufacturers' panels by code, because doing do voids UL listings (NEC 408).

29. SPD devices must be supplied with 14 AWG copper or aluminum (NEC 285.26).

30. **b.** NEC 410.130(F)(3) discusses thermal protection for luminaires installed in concrete.

31. **d.** The storage battery in a legally required standby system must source 87.5% of nominal voltage (NEC 701.11[A]).

32. In a series type of circuit, all current flows through all loads.

33. **b.** Grounding is unneeded for secondary circuits for lighting systems operating at 30 volts or less (NEC 411.5[A]).

34. **a.** Air conditioners and equipment with hermetic refrigeration are covered in NEC Article 440.

35. To ensure safety when wiring a hot panel, electricians should don safety glasses, wear leather gloves, and use tools rated for 1000 volts (*OSHA Handbook*).

36. **c.** NEC Table 210.24 summarizes branch circuit requirements.

37. A kitchen requires branch circuits (NEC 210.52[B][1]).

38. **a.** The temperature ratings of wires' outer jackets appear in NEC 310.13.

39. **c.** A continuous-duty power resistor conductor must be rated at 110% of full-load current (NEC Table 430.29).

40. **b.** Because wire gauge is based on fill rate, the number of wires that can fit into a box or conduit, fewer large-diameter wires can route through, or fill, a length of conduit. For example, 20 gauge wire is thinner than 10 gauge wire.

41. **c.** USE type conductors are not rated for a 90°C (194°F) maximum operating temperature (NEC Table 310.13[A]).

42. NEC Article 430, Part III contains the requirements for motor protection.

43. **b.** EMT can be used with exposed work (NEC 358.10 and 358.12).

44. **a.** NEC Table 310.16 is commonly used to temperature rate wires.

45. Hallways that are 10 ft. long require a receptacle at 6 ft. (NEC 210.52H).

46. **d.** Per NEC Section 215.2, electricians who are calculating transformer feeder ampacity should use transformers' nameplate ratings.

47. **a.** GFCI protection is needed for a hot-tub motor (NEC 680.71).

48. **a.** The interior of a raceway that is installed in a wet location is itself a wet location (NEC 300.9).

49. **a.** Electricians installing conductors in a raceway or cable should consult NEC Section 310.15(B)(2).

50. Per NEC 360.12, FMT type conduit and raceways are disallowed from battery storage areas.

51. **b.** ENT must be marked at intervals not to exceed 10 ft. (NEC 362.120).

52. A 100 ft. piece of RTRC that is cut at 20° will be 100 ft. 1.26 in. at 90° (NEC Table 355.44).

53. **b.** Intermediate IMC may serve as a grounding conductor (NEC 250.118[3]).

54. **c.** NEC Articles 517 and 660 both address x-ray machines; NEC Article 517 covers medical uses, NEC Article 660 concerns industrial x-ray installations.

55. **a.** Lightning protection system ground terminals should be bonded to the building or structure grounding electrode system (NEC 250.106).

56. **d.** Closets are small in volume and usually feature no ventilation (NEC 410.16).

57. According to NEC Section 200.6(B), a number 6 AWG or larger ground must carry a white or gray termination point.

58. **a.** A disconnecting means must exist "within sight" of a fixed space heater (NEC 424.19).

59. **b.** ENT must be supported every 3 ft. and at 3 ft. from a box (NEC 362.30[A]).

60. **a.** All new wiring receptacles must be grounded to a neutral bar in a panel box (NEC 250).

61. **c.** Single-throw knife switches must be installed so that gravity cannot close them (NEC 404.6[B]).

62. Direct-current motors are always protected at a minimum of 250% of the full load current (NEC Table 430.52).

63. The smallest residential service is 100 amps (NEC 230.79[C] and [D]).

64. **b.** A motor's starting current should form the basis for short-circuit and ground-fault protective devices (NEC 440.22[A] and 440.22[B]).

65. **b.** NEC requirements state that 120V track lighting should be mounted at least 5 ft. above the floor (NEC 410.151[C][8]).

66. **a.** Audible and visual signal devices, except portable standby power sources, should be provided, where practicable, for derangement and load carrying (NEC 702.7).

67. Per NEC 250.126(3), a hidden device terminal intended for a grounding conductor must be marked green, "GR," or "G" or carry a grounding symbol.

68. **a.** Air-conditioning and refrigeration motors are considered one machine, regardless of their locations (NEC 440.8).

69. **d.** NEC Article 342 covers IMC and describes its permitted uses.

70. **a.** The NEC has no voltage-drop requirements.

71. **c.** A 9 ft. countertop requires three receptacles (NEC 210.50).

72. **b.** Per NEC 424.44, heating elements are spaced at 1 in. minimum on-center.

73. Per NEC 285.4, one SPD can be used to protect ungrounded conductors when rated at 100V or less.

74. **b.** Electricians cannot install panels in breakers of different manufacturers, because doing so voids UL listings.

75. a. Per NEC Article 100, for ungrounded circuits, voltage to ground is defined as the greatest voltage between the given conductor and any other conductor of the circuit.

76. a. Generator locations must meet the NEC requirements set forth in Section 430.14 (NEC 445.10).

77. NEC Section 408.4 states that a circuit directory must be completed legibly so that it can be used by another electrician or understood by emergency personnel.

78. b. NEC Article 358 identifies EMT installation requirements.

79. For 12 AWG nonmetallic sheathed cable, size 12 ground wire applies, while 8 AWG nonmetallic sheathed cable takes size 10 ground wire.

80. a. Some dedicated-clock, undercabinet-microwave, and garbage-disposal receptacles can reside in kitchens lacking GFCI protection (NEC 210.8).

81. b. The door to an indoor transformer vault should be fire rated to 3 hours (NEC 450.43[A]).

82. d. The upper limit for Edison fuses is 30A (NEC 240.51).

83. c. Wire-binding screws or studs and nuts with upturned lugs or the equivalent can connect 10 AWG or smaller conductors (NEC 110.14[A]).

84. Surge arresters less than 1 kV are also known as Type 1 surge-protective devices, while transient voltage surge suppressors are known as Type 2 and Type 3 SPDs (NEC 285.1 FPN 1 and 2).

85. b. NEC Article 320 articles Type AC cable, while Type MC cable is the topic of NEC Article 330.

86. b. A flexible cord that supplies a room air conditioner should be no longer than 10 ft. for a nominal 120V rating or 6 ft. for a nominal 208 or 240V rating (NEC 440.64).

87. b. NEC 430.112 gives conditions under which motors may be served by a single disconnect.

88. a. When voltage across a resistance doubles and the resistance is halved, the current increases per Ohm's law.

89. The term *double insulated* indicates a chassis is isolated from a system ground (NEC 250.114).

90. a. Aluminum is unsuitable for grounding-electrode use, because it reacts with water (NEC 250).

91. b. When defining acceptable work for fire alarm systems, both NEC Articles 760.3 and 300 apply.

92. b. The common use of wire mold is protection of wires run on the surface of a commercial building.

93. a. Four residents is the minimum requirement for nursing-home operation (NEC 517.2).

94. NEC Section 314.28 outlines the requirements for sizing pull boxes.

95. b. There must be 30 in. between movie-theater projectors (NEC 540.12).

96. c. A new requirement in NEC 2008, receptacles above show windows must be within 18 in. of the windows' tops.

97. b. There is no such requirement in the NEC codebook.

98. d. Demand factors for elevator installations are based on a duty cycle of 50% (NEC Table 620.14 FPN).

99. Calculations for additional appliances appear in NEC Sections 220.16 through 220.20.

100. b. NEC considers a circuit that cannot cause a fire in a hazardous condition intrinsically safe.

101. The connection of electrical appliances is covered in NEC Article 422.

102. b. The National Fire Protection Association (NFPA) sponsors the NEC.

103. a. The conductors of a multiwire branch circuit must arise from the same panel (NEC 210.4).

104. Per Section 220.5(B), to make overall calculations correct, NEC rounds ampacity calculations downward.

105. NEC Section 440 discusses hermetic refrigerant and motors, motor circuits, and controllers.

106. c. The National Fire Prevention Association is represented by the acronym NFPA.

107. b. Porcelain-mounted receptacles require GFCI protection (NEC 210.52).

108. NEC Article 400 allows extension-cord use in wet areas.

109. a. Panelboard circuit directories are an NEC requirement (NEC 225.37).

110. Feeder cable installation requirements appear in NEC Article 215.

111. a. Unlike the other options, air conducts no electricity.

112. b. Junctions must be accessible, according to NEC 314.29.

113. Per NEC 422.16(B), a flexible cord may be used without a grounding conductor or termination when listed and marked as double insulated.

114. c. Per NEC 460.6(A), a capacitor must discharge to 50V or less within 1 minute.

115. a. Electric baseboard heaters cannot reside below receptacle outlets (NEC 422).

116. c. A cable management system is defined as an apparatus designed to control and organize unused lengths of cable or cord at electrified truck parking spaces (NEC 626.2).

117. To help ensure safety, NEC 404 requires that a light switch reside at each doorway.

118. b. An OCPD need not be inserted in series with a shunt winding during autotransformer installation (NEC 450.4).

119. d. Ceiling tiles cannot be used to support luminaires in any circumstances.

120. NEC Article 220, Part 2 covers computed loads and demand factors.

121. c. Because amperes is watts divided by volts, bulbs totaling 750 watts would have 6.25 amps at 120V (750 divided by 120).

122. a. NEC 250.28(A) covers the type of material electricians must use for grounding and bonding.

123. b. NEC Table 300.5 provides the requirements for buried cable of up to 600V.

124. b. The system bonding jumper is the difference between a grounded circuit conductor and an equipment grounding conductor at a separately derived system (NEC 250.2).

125. b. The number of switches on a set of lights is unlimited (NEC 404.2).

126. In a multiphase system in which one phase is used as a single phase source, the neutral conductor serves as ground (NEC 250.26[5] and [2]).

127. Current is measured in units known as amperes.

128. a. Switches and circuit breakers used as disconnects must indicate switch status (NEC 422.35).

129. **d.** Power equals voltage times resistance, which in this case is 2 × 12 or 24 watts.

130. NEC Section 250.24(C)(1) covers the routing and sizing of grounding conductors.

131. **d.** A ground ring around a building or structure, in direct contact with the Earth and consisting of at least 20 ft. of bare copper not smaller than #2 AWG, is an allowable electrode (NEC 250.52[A][4]).

132. **d.** Article 440 provides relevant definitions for AC and refrigeration equipment.

133. **b.** A surge arrester cannot be used in lieu of solid bonding or grounding. In fact, without solid bonding, a surge arrestor or suppressor is useless.

134. **a.** NEC Section 250.34 discusses portable generators, as well as those mounted on vehicles.

135. NEC Section 215(A)(3), FPN 2, recommends the limiting of voltage drops or feeders.

136. **b.** Per NEC Table C.1, 38 2/0 kcmil XHHW-2 conductors cannot be pulled into 4 in. trade size conduit.

137. In electrical terms, flexible metal conduit is known as Greenfield (NEC Article 360.2).

138. Using Ohm's law, the value of the fuse would be 10V/2 ohms, or 5 amperes.

139. **a.** Light switches are required at doorways (NEC 410.2).

140. **b.** The demand factors in NEC Table 220.12 do not apply when lighting is a concurrent load, such as in a restaurant.

141. **d.** The wire should be stapled 12 in. from the box, or per prevailing local code.

142. **b.** NEC Article 700 covers all aspects of hazardous locations, which NEC categorizes as "special occupancies."

143. **a.** NEC Article 225 outlines the fill requirements for conduit installed outdoors.

144. **a.** Gas hot water heaters require electricity to power their control circuitry.

145. **a.** NEC 240.50 through 240.61 outline fuse requirements; breaker requirements appear in 240.80 through 240.86.

146. When installing parallel conductors, electricians should consult NEC Section 310.4.

147. **b.** When run inside island cabinets, wire requires protection (NEC 224.10).

148. **b.** Wire that is 12 gauge takes 2.25 cubic inches (NEC 314.6[A]).

149. **b.** Elevators are discussed in NEC Section 620.21(A).

150. **d.** A box requires $\frac{1}{2}$ to $\frac{1}{4}$ in. outer jacket (NEC 314.17).

151. **b.** NEC does not authorize the use of the Earth as a bonding jumper, because the Earth's resistance is more than 100,000 times greater than that of a bonding jumper.

152. **d.** The NEC 2008 does not have a requirement for this, check with local code requirements.

153. NEC Article 390 outlines the requirements for installing raceways under floors.

154. **a.** Electricians may install lights in shower stalls (NEC 410.10[D]).

155. **c.** NEC 654.2 provides guidance on rooms designed for, but not functioning as, computer areas.

156. **a.** In a patient bed location in a critical care area, one panelboard should give rise to all branch circuits (NEC 517.18[A]).

157. A receptacle or switch accepting 14-2 counts as 4 cubic inches (NEC 314.16[B][4]).

158. According to NEC Section 422.31(B), a disconnect is not required when the breaker in a panel can be locked in the off position.

159. b. Regardless of whether GFCI protection is in place, a switch installed in a wet location must be rated for wet environments (NEC 404).

160. b. In NEC, Section 110.75 discusses manhole diameter.

161. Electrical requirements for unfinished basements appear in NEC Section 210.52(G)(1).

162. b. Because it is brittle at low temperatures and has a low melting point, solder is not allowed as a means of connecting grounding and bonding equipment (NEC 250.8[B]).

163. a. Per NEC 110.5, the default is copper unless otherwise stated.

164. a. NEC considers garage-door openers appliances (NEC 422.31).

165. b. Exposed ballasts cannot contact flammable material (NEC 410.136).

166. c. NEC 408.30 is the reference for choosing panelboards.

167. b. Although the NEC states that chemistry affects resistance, it does not advocate adding chemicals to do so (NEC 250.52[A][6]).

168. According to NEC Section 110.26(F)(1)(A), the zone, which equals the width and depth of equipment from floor to 6 ft. above the equipment, should be kept clear of all piping, ducts, or other equipment unrelated to the electrical installation.

169. a. A UL-listed product is adequate for its stated purpose and requires no further inspection (NEC 90.7).

170. b. Microwaves and garbage disposals should operate on separate circuits (NEC 220).

171. c. Electricians use NEC Table 450.3 to determine OCPD requirements for transformer primaries on the supply side.

172. Restrictions on the use of plastic junction or device boxes appear in NEC Article 314.3.

173. b. NEC 250.52(E) provides guidance for electricians installing supplementary electrodes.

174. b. Wire must extend only 6 in. outside a device box (NEC 301.14).

175. According to NEC Section 310.4, all the phase wires in parallel conductors must be the same length, type, and size.

176. b. NEC 250.30(A) prohibits an Earth connection on the load side of the grounding point of a separately derived system.

177. A grounding electrode conductor must be installed in one continuous length according to NEC Section 250.64(C).

178. a. (NEC Article 250.146).

179. c. Power wires should be 6 ft. from an attic access hole (NEC 320.23).

180. The requirements for thermal ballast protection appear in NEC Section 410.

181. d. A residence with four bedrooms on a second floor with one hall area, one bedroom on a first floor with two hall areas, and one basement would generally require nine smoke detectors for maximal safety (refer to prevailing local code).

182. NEC Article 410 covers the installation of lighting fixtures and supports.

183. a. NEC 410.7 outlines requirements for fluorescent ballasts that are installed indoors and require internal thermal protection.

184. **a.** The special requirements of installing lights in a clothes closet appear in NEC 410.8.

185. **b.** When a panel has 30 amps or fewer, its snap switches cannot exceed a 200-amp main breaker (NEC 408.36[A]).

186. **a.** According to NEC 110.3(B), breakers are acceptable when equipment is marked in the maximum fuse size.

187. **b.** According to NEC 517.39(C)(1), emergency circuits can sometimes run in the same box as normal wiring.

188. **a.** Per NEC 250.50, all ground conductors must be bonded.

189. **a.** When the lip of a box is recessed excessively in drywall, electricians may mount switches or outlet yokes directly to walls (NEC 314.20).

190. **d.** NEC's purpose, to provide for safety while using or installing electrical energy, appears in Section 90.1(A).

191. In reference to switchboards and panelboards, the phrase *condition one* means no live parts or terminations are exposed (NEC Article 110.26[A][1]).

192. NEC Section 240.6 contains the standard ampere rating for fuses and nonadjustable circuit breakers.

193. **b.** A grounded conductor termination bar should be insulated from the enclosure of a park trailer distribution panelboard (NEC 552.45[A]).

194. Examples of dedicated circuits in a residence include the furnace, sump pump, water heater, and dryer. Other examples include the clothes washer, spa, microwave, and refrigerator.

195. **c.** NEC 440.12(A)(1) and 440.12(B)(2) require that a disconnecting means be at least 115% of the sum of all loads in equipment.

196. **b.** NEC 440.52 addresses overload protection for motor compressor appliances.

197. **a.** A heavy-duty lampholder must be rated at 600W or greater (NEC 210.21[A]).

198. **b.** Per NEC Article 100, a bonding jumper insures a connection between metal parts and the earth.

199. **b.** When an appliance carries no breaker requirement, current rating may be derived through Watt's law (NEC 422.11[B] through [G]).

200. **d.** NEC 250.4(A)(5) stipulates that the earth is not the sole equipment grounding conductor.

201. NEC Table 5 contains information on the volumes of different conductors.

202. **b.** Per NEC 424.2, any fixed heater must be considered a continuous load (NEC 424.2[B]) for the purposes of branch circuit sizing, regardless of duty cycle.

203. **c.** The term *bundled* is defined in NEC Article 520.2.

204. The requirements for sizing main bonding jumpers appear in NEC Article 250.

205. **c.** The term *approved* means "acceptable to the AHJ" (NEC Article 100).

206. **b.** In a Class 2, Division 1, location, when a multiwire circuit's disconnecting device opens all ungrounded conductors simultaneously, a multiwire branch circuit is allowed (NEC 502.40).

207. NEC Section 517.13 requires two independent equipment grounding paths.

208. Non-current-carrying conductive materials enclosing electrical conductors or equipment are covered by NEC Section 250.4(A)(2).

209. **b.** The NEC states that two small-appliance circuits can not be shared (NEC 210.52[B][2]).

210. **a.** NEC 250.32(B) provides this requirement.

211. The authority having jurisdiction (AHJ) can order a building or residence disconnected from its power supply when any electrical equipment or its installation endangers human life or property (refer to prevailing local codes).

212. NEC Section 250.12 requires a clean surface to ensure a nonconductive coating for grounding or bonding.

213. **b.** NEC Article 312.3 disallows the installation of recessed boxes in combustible walls, specifying the relief in noncombustible walls.

214. **d.** A site-isolating device installed at a distribution point serving two or more agricultural buildings or structures should be permanently marked on the face of and immediately adjacent to the operating handle to identify its purpose (NEC 547.9[A][10]).7

215. Separately derived systems are covered under NEC Section 250.20(D).

216. All independent grounded neutral systems must be grounded according to NEC Sections 250.36 and 250.18(B).

217. **b.** Neutral current = square root of $(20^2 + 40^2 + 60^2 - [20 \times 40] - [40 \times 60] - [20 \times 60])$ = 35 amperes.

218. **b.** Ampacity is defined as the current, in amperes, a conductor can carry continuously under the conditions of use without exceeding its temperature rating (NEC 100).

219. **b.** NEC 250.24(A)(1) to (A)(5) provide reference material when a premises wiring system supplied by a grounded AC service must have a grounding electrode attached to a grounded service conductor.

220. **b.** Safety is the most important aspect when becoming a professional electrician.

221. This is the symbol for a television.

222. NEC 250.24(B) requires unspliced main bonding jumpers.

223. **b.** NEC 314.15 does not apply to overhead service installations.

224. **d.** Per NEC 215.2, which discusses derating factors, the feeder conductor calculation fails to consider temperature and other environmental factors.

225. **a.** Kitchen counter circuits can accommodate gas stoves (NEC 210.52[B]).

226. **a.** Per NEC Article 90, following the code promotes safety.

227. **a.** NEC 310.15(B)(2) establishes different temperature zones based on conduit's distance above roof level.

228. When using equipment bonding jumpers, electricians should reference NEC Section 250.102(C).

229. **c.** Using M/C cable in cupboards help safeguard wire.

230. **b.** A bathroom exhaust fan requires no GFCI protection.

231. Per NEC Article 100, an underground location is considered a wet environment.

232. **a.** Electrode-type boilers operating at over 600 volts, nominal, should be supplied only from a three-phase, four-wire, solidly grounded wye system or from isolating transformers arranged to provide such a system (NEC 490.71).

233. a. The phases of a three-phase panelboard must be arranged from left to right or from top to bottom when viewed from the front (NEC 408.3).

234. a. Seen most often in audio and data applications, crosstalk affects signals in other conductors.

235. a. Circuit breakers listed and marked as HID may serve to switch 120 and 277V fluorescent lighting circuits (NEC 240.83[D]).

236. Electricians may use wire ties to strap wire to a three-gang box.

237. From the floor, a main breaker should be 79 in. high (NEC 240.24[A]).

238. A wet niche light requires surrounding water to cool its bulb during operation.

239. b. According to fire codes, a separate circuit for smoke detectors risks fire damage or injury.

240. c. Per NEC 300.11, electricians may hang secondary supports from primary supports when running new work over old.

241. b. The neutral conductor of feeders supplying systems that use or may both phase-control and sine wave dimmers should not be considered current carrying (NEC 520.27[B][3]).

242. b. Bend radius is the tightness of a wire or conduit turn.

243. d. Edison-base plug fuses should be classified as not over 125 volts and 30 amperes (NEC 240.51[A]).

244. b. Wireway splices and taps must be accessible (NEC 376.56[A]).

245. c. According to NEC Table 5A, the diameter of a bare number 4 AWG conductor is 0.213 in.

246. b. Per NEC 517.18(A), with a few exceptions each patient bed must be supplied by at least two branch circuits to ensure redundancy.

247. a. ANSI, or the American National Standards Institute, develops and maintains standard symbology for prints and schematics.

248. The interior of any open or closed container of a flammable liquid should be considered a Class 1, Division 1, or Class 1, Zone 0, location (NEC 516.3[A][1]).

249. d. Per NEC 230.82, current limiting devices, photovoltaic systems, and instrumentation transformers are all disallowed from connection to the supply side of a service disconnect.

250. b. Per NEC 230.43 cablebus is one of the 16 specific types listed as a service entrance installation wiring method.

Appendix A Correlation Guide

This correlation guide will help you as you go through the Answers and Explanations section of this book. Each answer explanation provided for the Pretest, Chapter Reviews, and Posttests contains a cross reference to the NEC code. This guide breaks down the electrical code reference numbers not only by codebook topic, but also by chapter in this book.

Simply locate the code in the chart on pages 284–288 and read across to find the relating chapter, or section within a chapter, in this book. The first number in each code reference relates to the corresponding chapter in the NEC. For instance, if an answer provides a references to NEC 301.3, this can be found in Chapter 3 of the NEC.

CHAPTER CONTENT	NEC ARTICLE/CHAPTER
Basic Electrical Concepts (Introduction)	
Introducing Electricity Basics	Not in Codebook
Explaining Ohm's Law	Not in Codebook
Discussing Circuit Efficiency	Not in Codebook
Illustrating Transformer Action	Not in Codebook
Studying Fundamental Circuits	Not in Codebook
Learning Watt's Law	Not in Codebook
General Residential Requirements (Chapter 1)	
Understanding the Parts of Residential Installations	110.26 210.11, 220.12, 230.70 310.15(B)(6), 310.16, 314.17(C), 314.27, 314, 300.4, 334.15, 334.30 424.3,
Ensuring Protection in Residential Installations	210.18(A) and (B) 314.16
Working with Box Population and Fill	314.16
Grounding and Bonding Wire Properly	250.50, 250.56FPN, 250.52, 250.53(G), 250.64, 250.66,
In The Home: Room By Room (Chapter 2)	
Understanding General House Layout	210.52(A)(1-2), 210.50(A)(1-2), 210.52(H), 210.70(A), 210.70(A)(1), 250
Laying Out Bedroom Wiring	210.52(A)(1-2) 410.2
Running Wire in Hallways and Stairways	210.50(A)(1), 210.50(A)(2), 210.52(H), 210.70(A), 210.70(A)(1), 210.90(C)(2) 410.2
Wiring the Living Room	210.50(A)(1-2), 210, 210.50(A)(2), 210.70(A)(1),
Running Kitchen Wiring	210.8(4) thru (8), 210.11(C)(1), 210.19(A)(3), 210.20(D)(1), 210.50, 210.50(A), 210.50(B), 210.50(C)(1), 210.50(2), 210.52(B)(1), 210.52(B)(2), 210.52(B)(3), 210.52(C)(1), 210.52(3)(B) 410.2, 422.16, 422.16(A)(1)
Putting Utility Rooms Together	210.9(C)(2), 210.9(C)(3), 210.11(C)(2), 210.52(B), 218.70(3), 220.54 410.2, 410.10(D)

CHAPTER CONTENT	NEC ARTICLE/CHAPTER
Meeting Electrical Needs in the Bathroom	210.8(A), 210.11(C)(3), 210.50(D), 210.52(C)(1), 210.52(C)(1), 210.52(C)(1), 210.52(D), 210.70(A)(1), 410.10 (D), 410.90(D)
Installing Wire in Basements and Garages	210.20(C), 210.50 (1) and (2), 210.52(G)(1), 210.70(A)(2)(a), 210.70(A)(2)(b)
Placing Wiring outside the House	210.50(E), 210.50(E)(3), 210.52(E)(1)(2)(3), 210.63 320.23(A)
Residential Wiring Techniques (Chapter 3)	
Introducing Electricity Basics	Not in Codebook
Determining Service Requirements	114.3 210.2(B), 210.23(A), 210.52(C), 220, 220.12, 230, 250
Selecting the Correct Panel	110.3(B) 210.2, 220, Annex D
Creating the Service Entrance	230, 230.4, 250.5, 230.9, 230.9(A)(B), 230.22, 230.24, 230.27, 230.50, 230.54(F) 310.15(B)(6) 408.53 550
Connecting the Meter Base	230.5, 210.52
Harnessing the Power of the Panel	110.26(A)(2-3)(B)-(E) 220, 250.52(5-8) 314.29550 408.53 Annex D
Creating Wiring Infrastructure	110.1(E) 215.3, 220 300.4, 310.4(B), 314, 314.16(A)
Understanding Junction Boxes	215.11(E) 314.6, 314.21, 314.22, 314.23(1) and (2), 314.29
Running Interconnecting Wires	310.4(B), 310.15(B)(2), 334.30, 342.30, 360.2 422.17
Installing Three- and Four-Way Switches	210.21-53, 210.52 404.2
Exploring Switch Basics	404

CHAPTER CONTENT	NEC ARTICLE/CHAPTER
Identifying Proper Fixture Uses	100 210.8(B)(3) 314.27(A)(B) and (D) 701(A),
Applying Outside Fixtures, Boxes, and Switches	210.52(E) 376 408
Terminating Loads, Switches, and Receptacles	210.21, 210.52(A)(1), 210.70(A)(1)(2)(3) 310.4(B)
Routing and Dressing Cable	300.14, 310.4(B), 314.17, 314.17 A through D, 362.26,
Conducting Testing and Verification	Not in Codebook
Protection In The Home (Chapter 4)	
Understanding the Scope of Protection	250 340.10
Appreciating the Value of Fuses	Not in Codebook
Opting to Use Circuit Breakers	210.52(B)
Exercising Ground Fault Detection and Protection	210.8, 210.12, 210.52(C), 210.52(D), 210.52(G), 210.52(G)(1)(2), 210.52(E)(3) 422.52 680
Studying Arc Faults and Related Protection	210, 210.12(B), 210.52(H)
Using Smoke and Fire Detection	424.44(G)
Providing Personal Electrical Protection	Not in Codebook
General Commercial Electrical Requirements (Chapter 5)	
Differentiating Commercial Electrical Work	Not in Codebook
Identifying Unique Commercial Demands	90.2
Outlining the Scope of Transformers	Article 250, Part 2; 250.30 408.36, 450.21 through 450.28 505
Metering Power to Installations	230.71
Grounding Electrical Service	210.8(B), 250, 250.50, 250.64, 250.92, 250.104(A), 250.142(A) and (B)
Categorizing Commercial Feeder Requirements	110.14, 220, 220.60, 240

CHAPTER CONTENT	NEC ARTICLE/CHAPTER
Working with Commercial Panelboards	110.26 210.70(A)(1), 230.80, 230.62, 230.72(C) 408.30,
Identifying Characteristic Commercial Loads	310, 314.28
Working with Branch Circuit Wiring	314.12, 348.12, 360, 360.10, 500.3B
Working with Electrical Nonmetallic Conduit	362.10, 362.26 362.60, 362.120
Installing Weatherproofing	225.10
Navigating Explosive Areas	110.27(A) 250.118 300.4(H), 314.11, 314.23(H)(2), 358.30,
Wiring Health-Care Facilities	517.11, 517.20(a)
Commercial Wiring Techniques (Chapter 6)	
Calculating Electrical Loads	220, 220.14(F), 220.61
Mounting Panels and Boxes	300.4(B)(2)
Learning to Manipulate Conduit	215.2 356.10, 356.12, 356.20, 358
Using Fittings	
Applying Outlet and Receptacle Boxes	314.15, 314.27, 314.28
Installing Branch Circuits	250.118 344.10(C), 344.24-26, 348.20(A)(1), 348.60, 358.10, 358.30, 358.111, 360.11
Applying Rigid Nonmetallic Conduit	352.11-12 520.5, 518.4, 352, 362.11
Studying Conduit Bodies	300.15
Installing Branch Circuit Switches	100 210-20, 240.24 314, 314.25 404.8, 404.9, 290.5
Considering Special Switches	404.4, 406.8, 404.7, 409
Instituting Grounding in Raceways	250.50 300.20
Manipulating Grounding and Other Boxes	314, 392.3(A)

CHAPTER CONTENT	NEC ARTICLE/CHAPTER
Supporting Raceways	344.10 430
Completing Underground Installations	300
Accommodating Lighting and Illumination	250.119 400.2, 402, 402.5, 402.6
Posting Exit Signage	517.32(B), 517.42(B) 700, 710
Protection at the Office (Chapter 7)	
Understanding Circuit Basics	250, 250.122
Instituting Short-Circuit Protection	240
Instituting Overcurrent Protection	110 240
Rating Electrical Protection Devices	110
Classifying Fuses	240.6
Working with Circuit Breakers	240
Installing HVAC Applications	430.100, 450.46
Delivering Personal Safety and Protection	Not in Codebook
Industrial Applications (Chapter 8)	
Discussing Electrical Basics	Not in Codebook
Studying Electrical Control Systems	Not in Codebook
Looking to the Future	Not in Codebook

Appendix B Tables and Graphs

Scientific or mathematical endeavors often rely on graphs, tables, and charts to determine various types of data. This data, useful to varying degrees, may be presented in several formats.

A table or chart, one format for displaying data, lists values. While mathematical formulas often provide the same data as tables and charts, the latter often provide quicker routes to solutions. Figure A.1, for example, is a chart for converting Celsius and Fahrenheit temperatures. The far-right column quickly provides Fahrenheit temperatures, while the center column just as quickly lists Celsius equivalents.

When a target temperature lies between two known values, interpolation is needed. Interpolation involves determining where between two known values an unknown value lies. Assume we must convert 2.5 feet to inches, for example. Because 2.5 is halfway between 2 and 3, it makes sense that our answer should lie between 24 and 36 inches (see Figure A.2). The midpoint of 24 and 36 is the average of those two numbers, or 30.

MEASUREMENT	CELSIUS	FAHRENHEIT
Boiling point	100°C	212°F
Sweaty, hot weather	30+°C	85+°F
T-shirt and shorts weather	24°C	75°F
Average room temperature	21°C	70°F
Long-sleeve shirt and pants weather	15°C	60°F
Fleece jacket weather	10°C	50°F
Freezing	0°C	32°F

Figure A.1 Temperature Conversions

FEET	INCHES
1	12
2	24
3	36
4	48
5	60

Figure A.2 Converting Feet and Inches

Interpolation is easy when relationships are linear. The process is more complex when relationships are nonlinear. If, as Figure A.1 shows, 10°C is 50°F and temperature conversions were linear, 20°C should equal 100°F. It does not, however. Doubling Celsius temperatures does not double Fahrenheit equivalents; the relationship is nonlinear. Interpolation, therefore, is more of an art than a science.

For speed and accuracy, the NEC provides a great deal of data in table format. In most NEC tables, the first column contains the known or "independent" variables, while the second column lists "dependent variables," those values calculated using data in the first column. An electrician who knows the breaker size for an installation could consult Figure A.3 to determine what size wire to install.

Similarly, an electrician who knows wire size could consult Figure A.4 to identify which breaker to buy for a wiring installation.

BREAKER SIZE	WIRE SIZE
15	14
20	12
30	10
40	10
60	10
100	8
200	6
300	4
400	3
500	2
600	1
800	0
1000	00
1200	000
1600	0000
2000	250kcmil
2500	350kcmil
3000	400kcmil
4000	500kcmil
5000	700kcmil
6000	800kcmil

Figure A.3 Breaker and Wire Sizes (NEC Table 250-122)

WIRE SIZE	BREAKER SIZE
14	15
12	20
10	30
10	40
10	60
8	100
6	200
4	300
3	400
2	500
1	600
0	800
00	1000
000	1200
0000	1600
250kcmil	2000
350kcmil	2500
400kcmil	3000
500kcmil	4000
700kcmil	5000
800kcmil	6000

Figure A.4 Breaker and Wire Sizes Inverted

Figure A.5, another sample NEC table, has several columns referenced by wire gauge. This table is designed to show how many amperes can be specified for a particular wire gauge of a particular composition at a particular temperature rating. The table also reveals indirectly that certain conductor types are unsuitable for certain conditions. Type TW, for example, cannot be used in any ampacity at greater than 60° C. Either 14-gauge aluminum wire does not exist or it is disallowed.

In addition to providing wire-gauge information, Figure A.5 serves to help electricians research various options, such as what conductors can be used at 75° C to handle 25 amperes. The electrician should find 25s in the table and cross-reference the options:

12 Ga copper in RHW, XHHW, THW, SEU, THWN, USE or SER.

14 Ga copper in RHH, THHN, or XHHW

10 Ga aluminum in RHH, THHN, or XHHW

Tables are just one means of formatting data. The data values from tables can be plotted on horizontal and vertical axes to create graphs that show data trends as well as data points. While graphs are plotted using

	COPPER			ALUMINUM		
SIZE AWG	60° C TW NMC NM UF	75° C RHW XHHW THW SEU THWN USE SER	90° C RHH THHN XHHW	60° C TW NMC NM UF	75° C RHW XHHW THW SEU THWN USE SER	90° C RHH THHN XHHW
14	20†	20†	25†	—	—	—
12	25†	25†	30†	20†	20†	25†
10	30†	35†	40†	25†	30†	35†
8	40	50	55	30	40	45
6	55	65	75	40	50	60
4	70	85*	95*	55	65	75
3	85	100*	110*	65	75	85
2	95	115*	130*	75	90*	100*
1	110	130*	150*	85	100*	115*
0	125	150*	170*	100	120*	135*
00	145	175*	195*	115	135*	150*
000	165	200	225	130	155*	175*
0000	195	230	260	150	180*	205*

Figure A.5 **Ampacity Table**

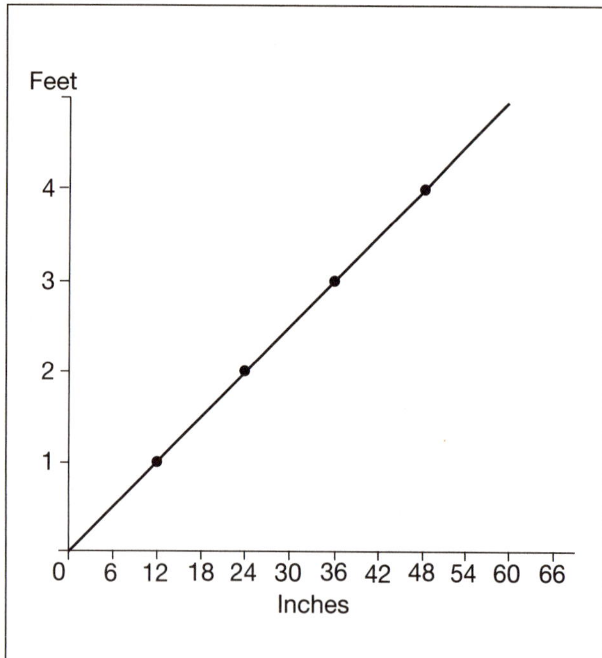

Figure A.6

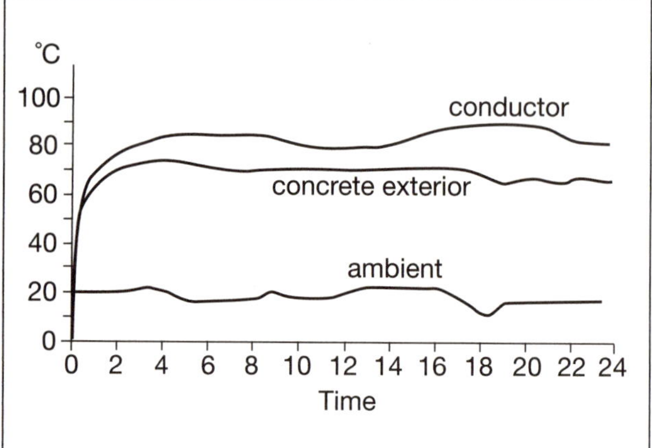

Figure A.7

two or three axes, the graphs in this appendix use only two axes. Figure A.6, one such graph, plots feet and inches in a straight line. Feet data points appear along the Y or vertical axis, while inch equivalents run along the X or horizontal plane. The 2-foot data point on the Y axis meets the line at the 24-inch mark on the X axis.

Interpolation applies to graphs as it does to tables. Recall that, using the table in Figure A.1, 2.5 feet interpolated to 30 inches. In the graph in Figure A.6, the 2.5-foot data point hits the line at 30 inches, verifying the result. Results are not always consistent in the two approaches, however. While interpolation applies to graphs, there is a higher likelihood of error related to the slope of the data line near the point of reference. Figure A.7, for example, has data lines with very little, almost flat, slopes.

Similarly, Figure A.8 has data lines with gentle slopes. As a result, value will deviate little and provide a narrow range of error. Figure A.9, in contrast, features a data line with a very steep slope. Here, even the slightest shift in the X direction noticeably impacts Y values. When electricians confront steep slopes, they can either (1) find more accurate data sources or (2) derate their reading, which means descending to the next-lower standard values.

In addition to tables and graphs, other useful data formats include nomographs. A nomograph is a chart that lists three or more values in columns. To read an unknown value in another column, a line intersects at least two constants.

Whatever the format, the data in all NEC charts and graphs undergo testing and verification. Accommodations are made for unique environmental conditions or other assumptions. A table listing wire gauges and ampacities, for instance, may assume an

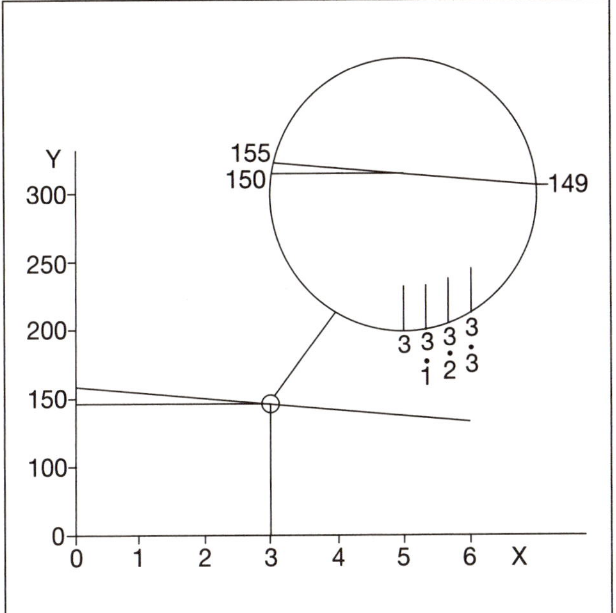

Figure A.8

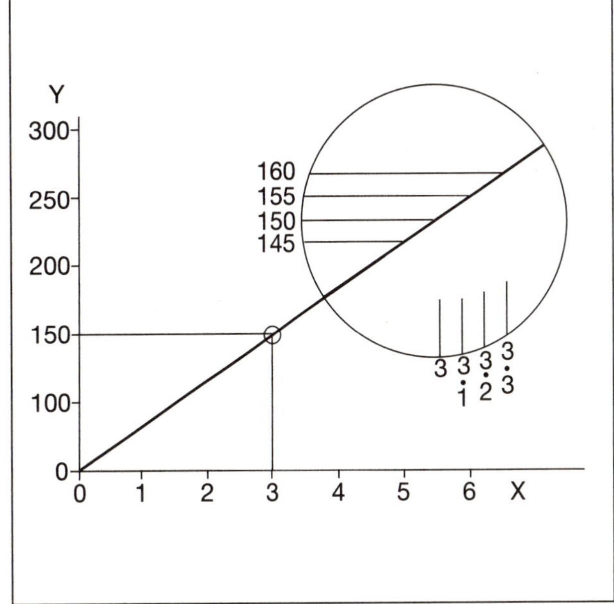

Figure A.9

ambient temperature of 70° F. When the temperature differs from 70° F, electricians must derate the referenced value.

Calculators can be a useful, time-saving tool when working with graphs, tables, and other data formats, but electricians should "estimate before they calculate" to help ensure accurate results. If, for example, an electrician is multiplying 325 and 97, which is close to multiplying 325 and 100, which yields 32,500, the result should be slightly less than 100.

Appendix C Electrician's Math

Electrical proficiency hinges on a grasp of arithmetic principles as well as an understanding of circuits, simple and complex. Series and parallel circuits, for example, operate on the theory that matter and energy are neither created nor destroyed. Simply put, whatever energy enters a circuit must exit. In addition, flow (current) increases as pressure (voltage) increases.

Completing Basic Calculations

According to Kirchoff's laws, all voltages and all currents must sum to zero. In a series circuit, all voltages sum to zero and the sum of component voltages equals the source voltage. In a parallel circuit, all currents sum to zero and the sum of component currents equals the source current. Ohm's law helps determine circuit current, while Watt's law calculates circuit power.

Consider Ohm's law, which states that the current through a conductor is both directly proportional to the potential difference, or voltage, across two points and inversely proportional to the resistance to those points. Written mathematically, Ohm's law looks like the following:

$$I \text{ (current)} = V \text{ (voltage)} / R \text{ (resistance)}$$

To solve for the other two variables—voltage (V) or resistance (R)—Ohm's equation can be rewritten as follows:

$$V = I \times R$$

$$R = V / I$$

The quantity to the left of the equals sign—the unknown—is the "dependent" variable, because its value depends on the other, "independent" quantities—the quantities we know. The quantities we know determine which equation best applies. Assume, for example, that we know resistance (R) is 32 ohms and voltage (V) is 120. If we chose to use the equation $R = V / I$, our calculation would look like the following:

$$32 = 120 / I$$

To determine our unknown quantity, I, we must express I as a function of V and R so that I is on one side of the equals sign and V and R are on the other. What we do to one side—add, subtract, multiply, or divide—we must do to the other to maintain balance between the two sides. Using our example, then, to balance the two sides of our equation and isolate the quantity I on one side, we would do the following:

$$R = V / I$$

Multiply both sides by I

$$R \times I = V \times I / I; I / I = 1$$

$$R \times I = V$$

Multiply both sides by 1 / R

$$R \times I / R = V \times 1 / R$$

$$I = V / R$$

▶ Applying Basic Math to Series Circuits

Basic math calculations are key to understanding the series and parallel circuits that are at the heart of electrical work. The following method allows you to track your results and progress and solve as much as you can with the least amount of work and the highest degree of certainty. Consider the series circuit in Figure B.1:

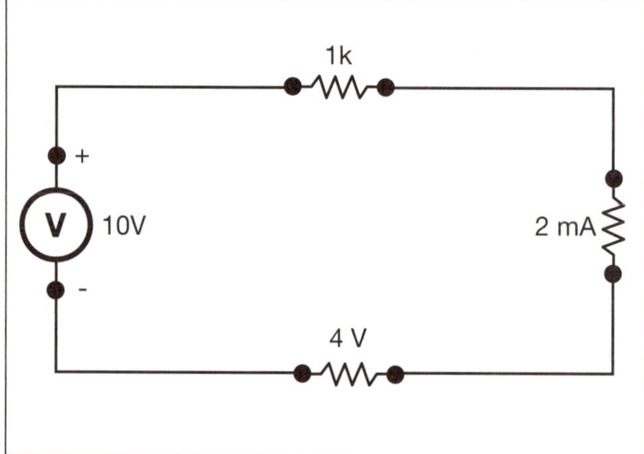

Figure B.1

To determine the voltage, current, and resistance for each component, as well as the total resistance, voltage, and current, we will use the following table:

COMPONENT	VOLTAGE	CURRENT	RESISTANCE
R1			
R2			
R3			
Total			

First, record the values that can be determined through inspection. Examine the circuit to see what you know, what you are given, and which independent mathematic variables are available.

COMPONENT	VOLTAGE	CURRENT	RESISTANCE
R1			1K
R2		2 mA	
R3	4V		
Total	10V		

In this example, only one value is known for each component in the circuit, and only one value is known for the total circuit. Recall, however, that Kirchoff's current law for series circuits states that the current in R_2, which we know, must be the same as the current in R_1, because all current leaving R_1 flows into R_2. The current in a series circuit must be the same in each component, and it must equal that supplied by the source.

Therefore, transfer the R_2 value to R_1, R_3, and total as shown in the following:

COMPONENT	VOLTAGE	CURRENT	RESISTANCE
R1		**2 mA**	1K
R2		2 mA	
R3	4V	**2 mA**	
Total	10V	**2 mA**	

The "Total" row now has two of the three variables we need to calculate the third (resistance) using Ohm's law:

$$R_{total} = V / I = 10V / 2mA = 5\ K$$

COMPONENT	VOLTAGE	CURRENT	RESISTANCE
R1		2 mA	1K
R2		2 mA	
R3	4V	2 mA	
Total	10V	2 mA	**5K**

Similarly, the "R_3" row has two of the three variables we need, so solve for resistance (R):

$$R_3 = 4V / 2mA = 2K$$

COMPONENT	VOLTAGE	CURRENT	RESISTANCE
R1		2 mA	1K
R2		2 mA	
R3	4V	2 mA	**2K**
Total	10V	2 mA	5K

Now, calculate R_2 using the idea that series resistances add. (Note: We could instead solve for V across R_1 first.)

$$R_{total} = R_1 + R_2 + R_3$$

$$R_2 = (R_{total} - R_1 - R_3)$$

$$R_2 = (5K - 1K - 2K) = 2K$$

Transferred to the table, this information appears as follows:

COMPONENT	VOLTAGE	CURRENT	RESISTANCE
R1		2 mA	1K
R2		2 mA	**2K**
R3	4V	2 mA	2K
Total	10V	2 mA	5K

To calculate values for the remaining fields—R_1 and R_2—do the following:

$$V_{R1} = I \times R = 2mA \times 1K = 2V$$

$V_{R2} = I \times R = 2mA \times 2K = 4V$ (Note: This is the computation for R3 earlier. Same values, same result.)

Using this information, complete the table:

COMPONENT	VOLTAGE	CURRENT	RESISTANCE
R1	**2V**	2 mA	1K
R2	**4V**	2 mA	2K
R3	4V	2 mA	2K
Total	10V	2 mA	5K

To ensure accuracy, complete a final calculation. We know from series-circuit theory that the voltages across each resistor, when added, must equal the total voltage, or 10V in this example. When the voltages across R_1, R_2, and R_3 are added (2V + 4V + 4V), they equal 10V, so our computation is sound.

▶ Calculating Values for Parallel Circuits

The computation methods for parallel circuits differ from, but are complementary to, those for the series circuits preceding. Consider Figure B.2, which shows a parallel circuit:

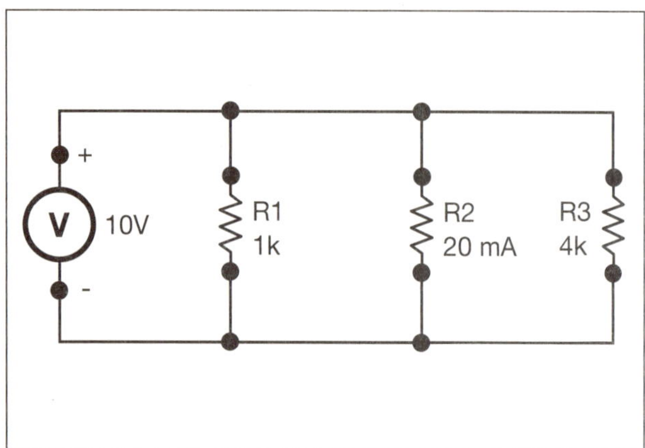

Figure B.2

As was done in the series circuit, make a table with spaces for all components and their values:

COMPONENT	VOLTAGE	CURRENT	RESISTANCE
R1			
R2			
R3			
Total			

Enter the values we can gain from the figure:

COMPONENT	VOLTAGE	CURRENT	RESISTANCE
R1			1K
R2		20 mA	
R3			4K
Total	10V		

Although no rows have two of three variables known, we know that, in parallel circuits like this one, the voltages across each component equals the source voltage. Update the table accordingly as follows:

COMPONENT	VOLTAGE	CURRENT	RESISTANCE
R1	**10V**		1K
R2	**10V**	20 mA	
R3	**10V**		4K
Total	10V		

Now we can begin calculating unknowns in any desired order. Working from top to bottom starting with R1, the following calculation applies:

$$I = V / R = 10V / 1K = 10mA$$

COMPONENT	VOLTAGE	CURRENT	RESISTANCE
R1	10V	**10 mA**	1K
R2	10V	20 mA	
R3	10V		4K
Total	10V		

Use Ohm's law to compute the value for R2:

R2 = V / I = 10V / 20mA = 500 ohms

COMPONENT	VOLTAGE	CURRENT	RESISTANCE
R1	10V	10 mA	1K
R2	10V	20 mA	**500 Ohms**
R3	10V		4K
Total	10V		

Next, calculate the current through R3:

I = V / R = 10V / 4K = 2.5mA

COMPONENT	VOLTAGE	CURRENT	RESISTANCE
R1	10V	10 mA	1K
R2	10V	20 mA	500 Ohms
R3	10V	**2.5 mA**	4K
Total	10V		

Now, apply Kirchoff's current law to determine the total current:

$$I_1 + I_2 + I_3 = I_{total} \text{ or } 10mA + 20mA + 2.5mA = 32.5mA$$

COMPONENT	VOLTAGE	CURRENT	RESISTANCE
R1	10V	10 mA	1K
R2	10V	20 mA	500 Ohms
R3	10V	2.5 mA	4K
Total	10V	**32.5 mA**	

Last, calculate total resistance in two ways: once to identify a value and a second time to verify the answer. First, calculate a value as follows:

$$R_{total} = V_{total} / I_{total} = 10V/32.5mA \approx 307.69 \text{ ohms}$$

COMPONENT	VOLTAGE	CURRENT	RESISTANCE
R1	10V	10 mA	1K
R2	10V	20 mA	500 Ohms
R3	10V	2.5 mA	4K
Total	10V	32.5 mA	**307 Ohms**

The 307 ohms value makes sense, because we know that the total resistance of a parallel circuit must be less than the lowest branch resistance in the circuit. In this circuit, that is 500 ohms.

Conduct a second calculation using the parallel resistance formula, which should further validate our answer:

$$Rt = \frac{1}{1/R_1 + 1/R_2 + 1/R_3}$$

Rt = 307.69 ohms

▶ Solving Equations for Combination Circuits

The mathematical principles that apply to series and parallel circuit analysis apply as well to combination circuits: circuits that use series and parallel elements together. There is no consistent way to solve combination circuits, however, because each circuit is unique and thereby requires a slightly different solution. Consider the combination circuit in Figure B.3:

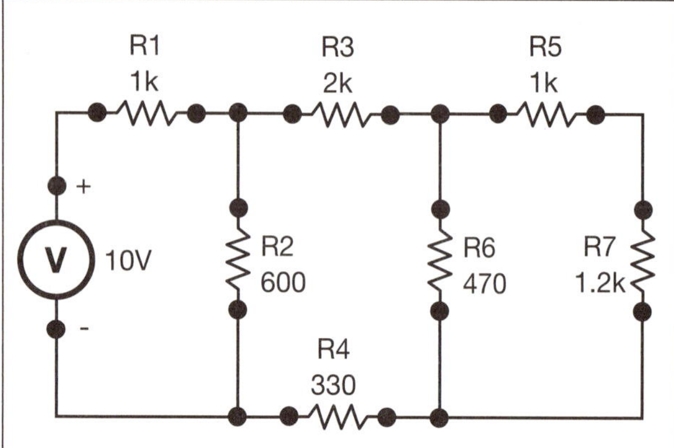

Figure B.3

Begin the computation by noting each component's voltage polarities as shown in Figure B.4:

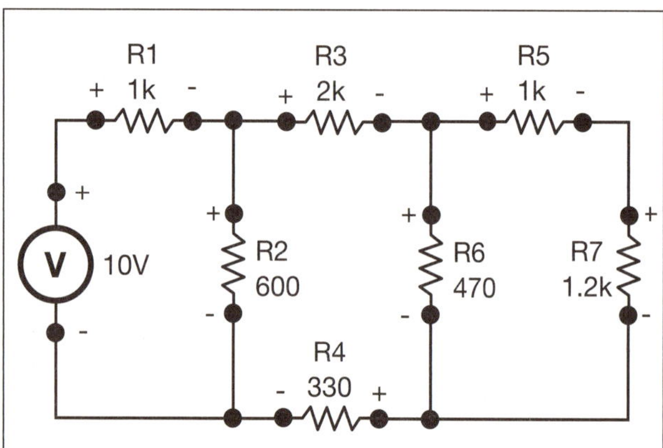

Figure B.4

Next, create a table like those in previous examples:

COMPONENT	VOLTAGE +	VOLTAGE –	V	R	I
R1					
R2					
R3					
R4					
R5					
R6					
R7					
Total					

This table requires additional entries, because as voltages around the circuit are determined, we must often calculate the voltage across a component by comparing (subtracting) the individual voltages on either side of that component. In the table, enter the values given in the circuit:

COMPONENT	VOLTAGE +	VOLTAGE –	V	R	I
R1				1K	
R2				600	
R3				2K	
R4				330	
R5				1K	
R6				470	
R7				1200	
Total	10	0	10V		

According to the figure, all components tied directly to the source have one known voltage: 0V or 10V. Enter these values in the table as follows:

COMPONENT	VOLTAGE +	VOLTAGE –	V	R	I
R1	10V			1K	
R2		0V		600	
R3				2K	
R4		0V		330	
R5				1K	
R6				470	
R7				1200	
Total	10	0	10V		

To calculate the total circuit resistance for a combination circuit, use series and parallel resistance formulae. Doing so often involves redrawing the circuit numerous times to reduce it to one resistance, but then Ohm's law can be used to calculate total circuit current, which is needed to progress the analysis. Record the steps you take. Once current is established, we will return the circuit to its original form to determine the values for each component.

Let's redraw the circuit, each time slightly different, in order to reduce it to one resistor.

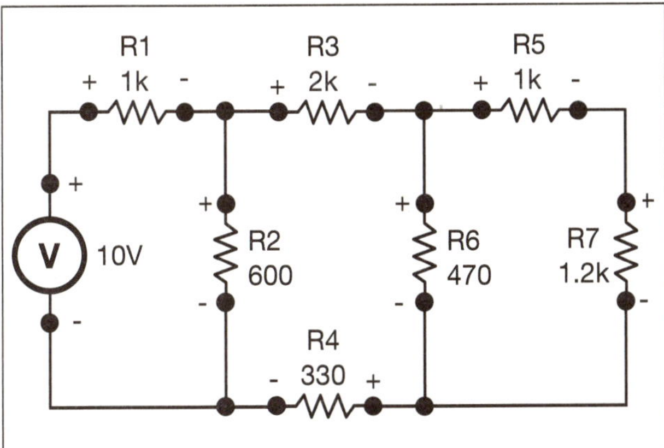

Figure B.5

Combine R5 and R7, in series (add):

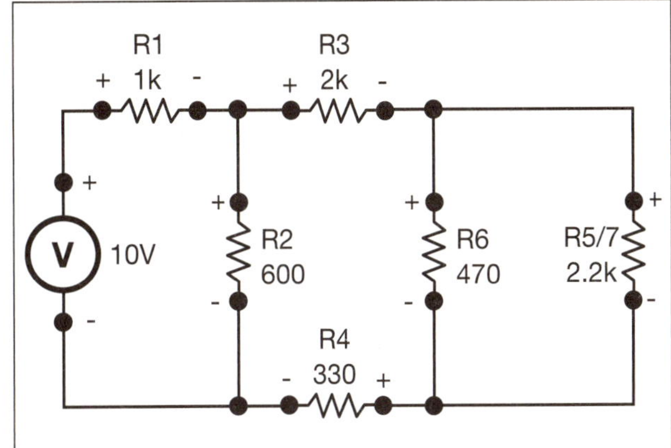

Figure B.6

Combine R6 with the result of the last combination, in parallel (reciprocal):

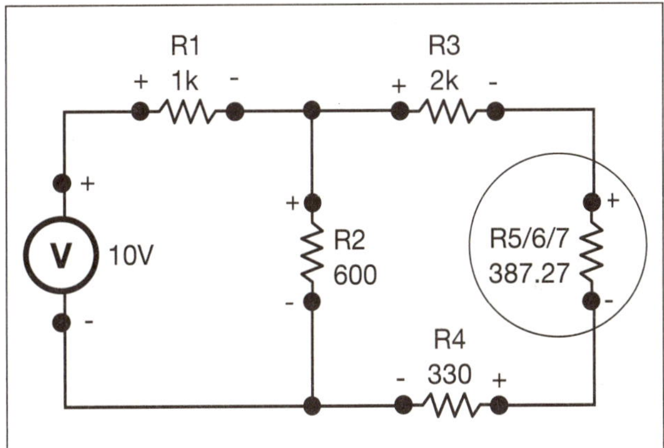

Figure B.7

Combine this result with R4 in series (add):

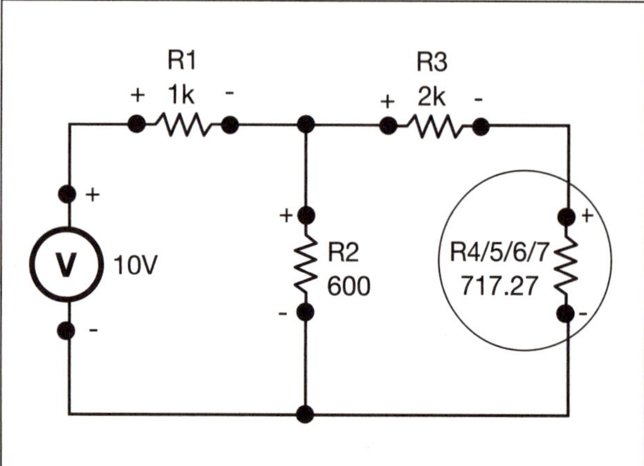

Figure B.8

Combine this result with R3 in series (add):

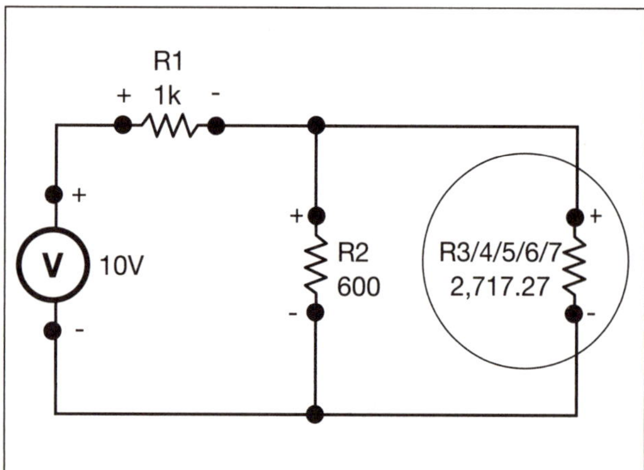

Figure B.9

Combine this result with R2 in parallel (reciprocal):

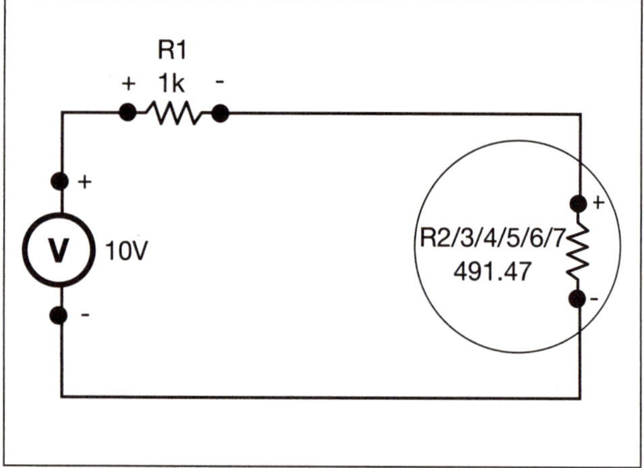

Figure B.10

Combine this with R1 in series (add):

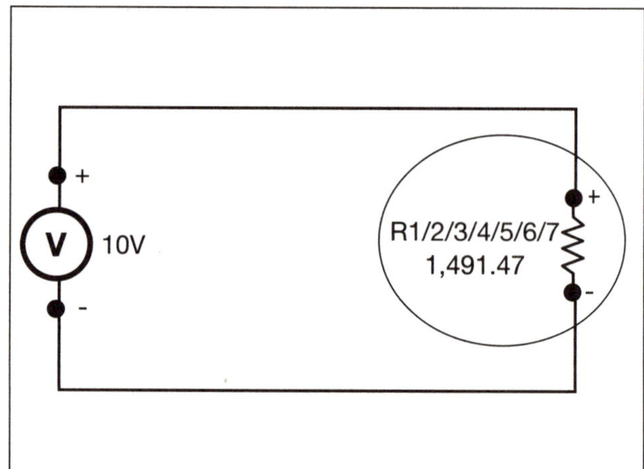

Figure B.11

Now, we are down to one resistance. Total current is equal to total voltage / total resistance, or in this case, 6.70476 mA or 6,704.76 microamperes.

Enter that in the table, along with the total resistance.

COMPO-NENT	VOLTAGE +	VOLTAGE –	V	R	I
R1	10V			1K	
R2		0V		600	
R3				2K	
R4		0V		330	
R5				1K	
R6				470	
R7				1200	
Total	10	0	10V	**1,491.47**	**6,704.76uA**

Returning to Figure B.10:

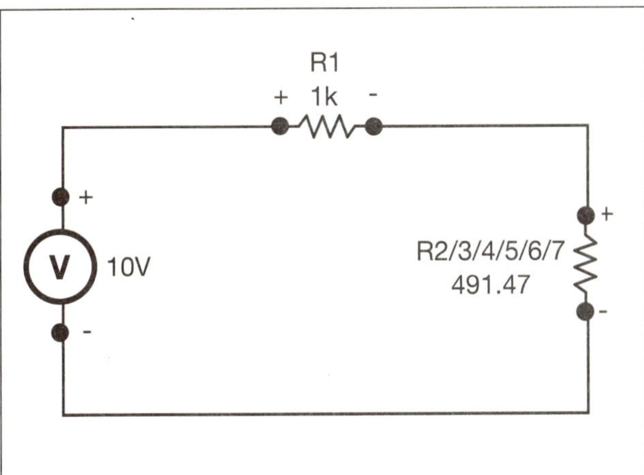

Now, using our total current value of 6,704.76 uA, we can calculate the voltages across the two components shown in the circuit since they are in series.

VR1 = 6704.76uA × 1000 Ohms = 6.70476 Volts

The voltage across R1 is 6.70476 Volts, and we know that the voltage on the + side of R1 is 10V, therefore the voltage on the – side of R1 must be 10V – 6.70476V or 3.295V. Enter these in the table:

COMPO-NENT	VOLTAGE +	VOLTAGE –	V	R	I
R1	10V	**3.295V**	**7.70476**	1K	**6,704.76uA**
R2		0V		600	
R3				2K	
R4		0V		330	
R5				1K	
R6				470	
R7				1200	
Total	10	0	10V	1,491.47	6,704.76uA

Now, review Figure B.9:

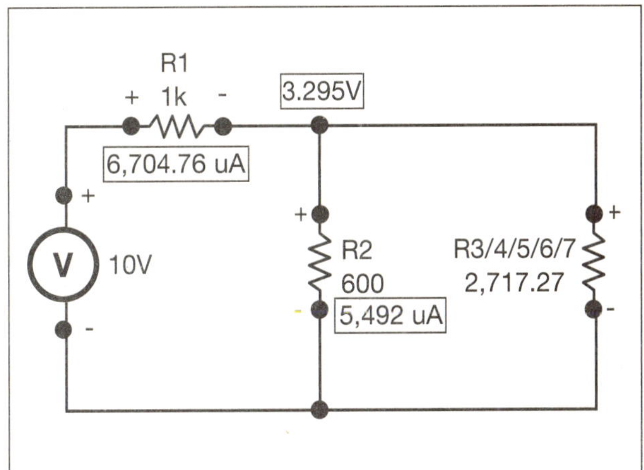

Placing the 3.294V we have determined in the previous step, we see that the same voltage appears at the + side of R2. Enter that in the table:

COMPO-NENT	VOLTAGE +	VOLTAGE –	V	R	I
R1	10V	3.295V	7.70476	1K	6,704.76uA
R2	**3.295V**	0V	**3.295V**	600	**5,492uA**
R3				2K	
R4		0V		330	
R5				1K	
R6				470	
R7				1200	
Total	10	0	10V	1,491.47	6,704.76uA

We see from this entry that we can now compute the voltage across R2 as 3.295V – 0V = 3.295V. Current in R2 is 3.295V/600 Ohms or 5.492 mA. Kirchoff's Current Law tells us that the current flowing through the combination of R3/4/5/6/7 is what's left over, or 6704.76uA – 5492uA or 1213uA.

Look at Figure B.8:

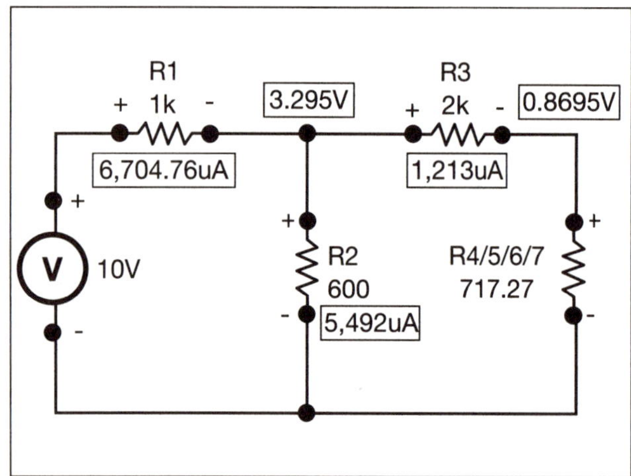

And, the table is now revised also:

COMPO-NENT	VOLTAGE +	VOLTAGE –	V	R	I
R1	10V	3.295V	7.70476	1K	6,704.76uA
R2	3.295V	0V	3.295V	600	5,492uA
R3	**3.295V**	**0.8695V**	**2.4255V**	2K	**1,213uA**
R4		0V		330	
R5				1K	
R6				470	
R7				1200	
Total	10	0	10V	1,491.47	6,704.76uA

Now that we have two out of three values for R3, we can calculate the voltage across R3 using Ohm's Law:

$$VR3 = 1213uA \times 2000 \text{ Ohms} = 2.4255V.$$

The voltage at the Minus side of R3 is the difference between the voltage + and voltage across, or 3.295V – 2.4255V which is 0.8695V.

We've done all we can for now; let's look at Figure B.7 again.

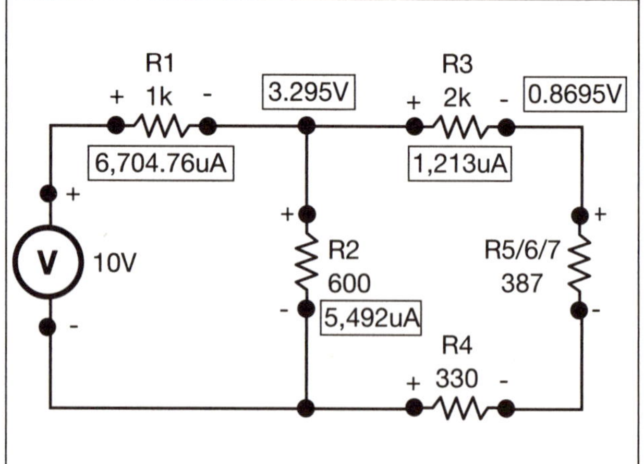

Note that we must also have 1213uA flowing through R4 (Kirchoff's Current Law) and that this value gives us two out of three R4 values and a voltage across R4 of 330 ohms × 1213uA (Ohm's Law) or 0.4003 volts can be calculated.

The table now looks like this:

COMPO-NENT	VOLTAGE +	VOLTAGE –	V	R	I
R1	10V	3.295V	7.70476	1K	6,704.76uA
R2	3.295V	0V	3.295V	600	5,492uA
R3	3.295V	0.8695V	2.4255V	2K	1,213uA
R4	**0.4003V**	0V	**0.4003V**	330	**1,213uA**
R5				1K	
R6				470	
R7				1200	
Total	10	0	10V	1,491.47	6,704.76uA

Also, we can calculate the voltage on the + side of R4 by adding the component voltage of 0.4003V to the – side voltage of 0V for a + side voltage of 0.4003V.

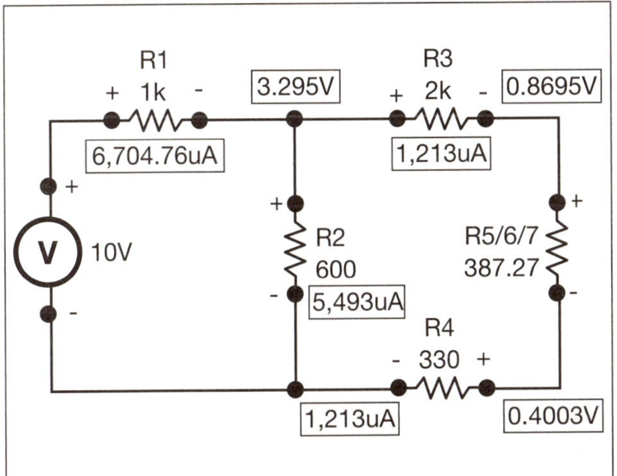

Figure B.6 now looks like this:

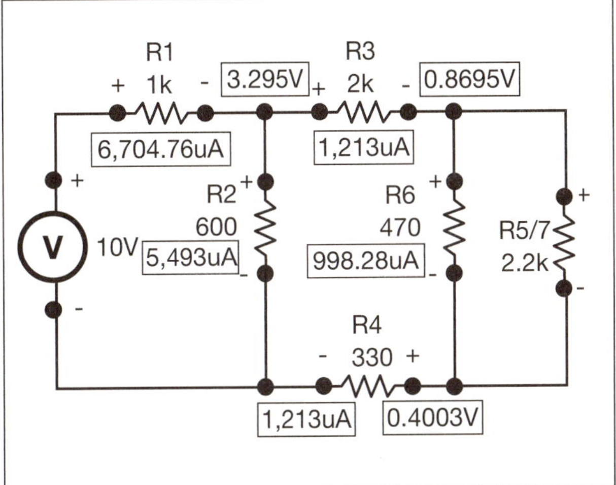

We can see by inspection that we have voltages for the + and – sides of R6, and we can easily write the component voltage as the difference between these two values, or 0.4692V.

Use this information to update the table:

COMPO-NENT	VOLTAGE +	VOLTAGE –	V	R	I
R1	10V	3.295V	7.70476	1K	6,704.76uA
R2	3.295V	0V	3.295V	600	5,492uA
R3	3.295V	0.8695V	2.4255V	2K	1,213uA
R4	0.4003V	0V	0.4003V	330	1,213uA
R5				1K	
R6	**0.8695V**	**0.4003V**	**0.4692V**	470	
R7				1200	
Total	10	0	10V	1,491.47	6,704.76uA

It is apparent that we have two out of three unknowns to solve Ohm's Law for R6.

IR6 = 0.4692V/470 Ohms = 998.28uA

COMPO-NENT	VOLTAGE +	VOLTAGE –	V	R	I
R1	10V	3.295V	7.70476	1K	6,704.76uA
R2	3.295V	0V	3.295V	600	5,492uA
R3	3.295V	0.8695V	2.4255V	2K	1,213uA
R4	0.4003V	0V	0.4003V	330	1,213uA
R5				1K	
R6	0.8695V	0.4003V	0.4692V	470	**998.298uA**
R7				1200	
Total	10	0	10V	1,491.47	6,704.76uA

Move on to Figure B.5, which is our original circuit:

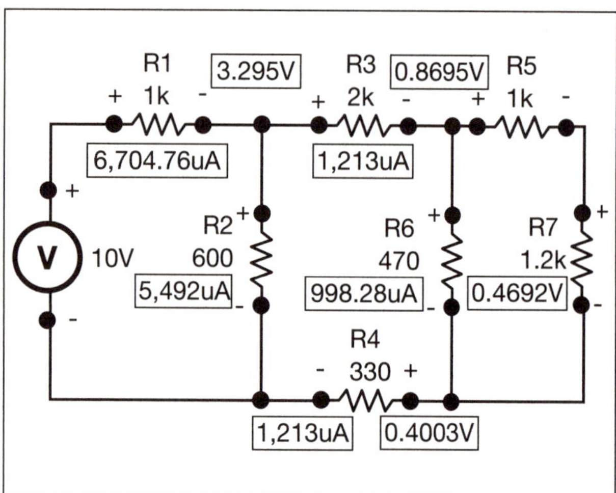

From Kirchoff's Current Law, it appears that the current through R5 and R7 (same value as they are in series) will be what's left over from the 1213uA after 998.298uA have taken the path through R6. This amount will be equal to 214.7uA. Add this to the table:

COMPO-NENT	VOLTAGE +	VOLTAGE –	V	R	I
R1	10V	3.295V	7.70476	1K	6,704.76uA
R2	3.295V	0V	3.295V	600	5,492uA
R3	3.295V	0.8695V	2.4255V	2K	1,213uA
R4	0.4003V	0V	0.4003V	330	1,213uA
R5				1K	**214.7uA**
R6	0.8695V	0.4003V	0.4692V	470	998.298uA
R7				1200	**214.7uA**
Total	10	0	10V	1,491.47	6,704.76uA

This gives us two out of three of our unknowns for both R5 and R7; we can now calculate the respective voltages across each and the + side and – side voltages for R5 and R7, and we are done!

VR5 : 214.7uA × 1000 Ohms = 214.7 mV

VR7 : 214.7 uA × 1200 Ohms = 257.6 mV

COMPO-NENT	VOLTAGE +	VOLTAGE –	V	R	I
R1	10V	3.295V	7.70476	1K	6,704.76uA
R2	3.295V	0V	3.295V	600	5,492uA
R3	3.295V	0.8695V	2.4255V	2K	1,213uA
R4	0.4003V	0V	0.4003V	330	1,213uA
R5	**0.8695V**	**0.6548V**	**0.2147V**	1K	214.7uA
R6	0.8695V	0.4003V	0.4692V	470	998.298uA
R7	**0.6579V**	**0.4003V**	**0.2576V**	1200	214.7uA
Total	10	0	10V	1,491.47	6,704.76uA

The table, and the analysis, are complete. Some of the additive values do not add exactly to the sum given; this is due to rounding off numbers as calculations are performed. Keep in mind that although these inaccuracies exist (and they will in nearly all calculations) they are overshadowed by the tolerances of the components, power supply, and measurement instruments.

Again, it must be stated that the specific steps in this example pertain to this example. Every circuit configuration will require a different combination or sequence of steps, but remember that the "tools" we have used in these analyses are:

- Ohm's Law
- Kirchoff's Voltage Law
- Kirchoff's Current Law

The difference between the analyses of different circuits lies in the order in which specific steps are taken. It will take some practice for you to be able to recognize where to start, what step to take in what order, and what info is "hidden" in the schematic or in the table.

It may help to mentally picture these circuits and their analyses as puzzles rather than as problems. It will also help you to recognize instances of Kirchoff's Law and use them to check the work you have done, as we did in the series analysis and the parallel analysis.

Like many analyses in the field of electronics, developing a methodical step-by-step procedure is usually of benefit to you, and in most cases over time you rely less and less on the published procedure and more on your experience in order to solve these circuits.

In summary, the ability to recognize, analyze, and troubleshoot series and parallel circuitry is fundamental to a career in electronics. After all, any circuit, no matter how simple or complex, can be broken down into series and parallel components to which Kirchoff's Laws and Ohm's Law, as well as Watt's Law can be applied.

Understanding the Math of Reactive Loads

So far, we have only looked at one type of load—a resistive load. There are also loads that are referred to as "reactive" loads; these are loads that store energy, like capacitors and inductors. The presence of these items and their ability to store energy requires that we adjust our calculations a little, and perform a few new ones.

To illustrate this, let's look at a common garden hose with a trigger nozzle. The hose holds 10 gallons of water and this water stays in the hose when the water is turned off after filling a five gallon bucket. After a few minutes, the water left in the hose leaks out and soaks into the ground.

Let's look at what just happened.

Amount of water in the bucket:	7 gallons
Amount of water left in the hose:	10 gallons
Amount of water through the meter:	17 gallons
Cost per gallon from water company:	\$0.15
Water used	7 × \$0.15 = \$1.05
Water wasted:	10 × \$0.15 = \$1.50
Total amount paid:	17 × \$0.15 = \$2.55
Efficiency factor:	\$1.05/\$2.55 = 0.41, or 41%

From a consumer perspective, we just paid \$2.55 for \$1.05 worth of water.

In your backyard the above scenario may happen once a day. In an electrical circuit, it happens sixty times per second! While the calculations for electrical power are somewhat different from, and more complex than the simplified model given, the concept is similar and easier to grasp.

At this point, it will be quite helpful to have a scientific or electrician's calculator available, as we will be calling on our dear friend Pythagoras as well as his sidekicks sine and cosine. These calculations are a bit too robust for most mental gymnasts, never mind us regular folks.

In the electrical field, capacitors, which store voltage (like our garden hose–stored water) are generally seen as the contributors of reactive power. Inductors, which are an opposing form of reactance, work as acceptors of reactive power; in fact, in a few pages, we will use an inductor to counteract the effects of capacitance in our circuit.

Consider a circuit with a capacitor in parallel with a resistive load. Start by calculating the reactance of the capacitor. Reactance, by definition, depends on the frequency as well as the capacitance value, the formula we use for this calculation is:

$$\text{Capacitive Reactance (Xc)} = \frac{1}{2\pi f C}$$

F is the frequency (60 Hz)

C is the Capacitance, in Farads (100 uF)

$$\text{In our case, assume Xc} = \frac{1}{6.28 \times 60 \times .001} \approx 27 \text{ ohms.}$$

Next, determine the reactive current through Ohm's law:

$$I = V / Xc = 120 \text{ volts}/27 \text{ ohms} = 4.44 \text{ amperes}$$

The reactive power is then:

$$4.44\text{A} \times 120\text{V or } 533.33 \text{ W}$$

Now, for the resistive current via Ohm's law:

$$I = V / R = 120\text{V} / 15 \text{ ohms} = 8 \text{ Amperes}$$

The resistive power is then:

$$8 \text{ Amperes} \times 120\text{V} = 960\text{W}$$

Remember that these two power values do not happen at the same time. The hose we discussed did not begin to exert pressure until it was full, which happened after the water was applied. In an electrical circuit, which uses a sinusoidal (circular) waveform, the voltage in the capacitor does not become evident until the source voltage reaches its maximum, which implies a 90° phase shift.

When we look at the relationship between reactive power and resistive power, we can plot them at 90° (right) angles on a graph. Now we must bring in the Pythagorean theorem. The actual power taken into the circuit is a combination of, but not a sum of, the two component powers. We must do the following calculation, based on a right triangle, to get the actual power we buy from the power company.

$$\text{Total Power} = \sqrt{(\text{reactive power})^2 + (\text{resistive power})^2}$$

$$\text{Total Power} = \sqrt{(533.33)^2 + (960)^2} = 108 \text{ Watts}$$

The power factor is then:

$$\frac{960\,W}{1098\,W} = 0.87$$

In other words, you are getting work out of 87% of the energy you are buying. We can correct for this low power factor by, curiously enough, adding a component to our circuit. The component we will add will allow us to store the energy left over rather than waste it, and we can then use that stored energy to do work.

We will add an inductor, so that the reactance of the inductor is the same as the reactance of the capacitor. They will then cancel each other out. As stated, we will need the inductor to have a reactance of 27 ohms—the same as our capacitor.

Using the inductive reactance formula:

$$\text{Inductive Reactance} = \sqrt{2\pi \times \text{frequency} \times \text{inductance}}$$

We will rearrange the formula to solve for the inductance, L:

$$L = \frac{Xl^2}{2\pi f}$$

Plugging in values, yields

$$L = \frac{27^2}{2\pi(60)}$$

Or, L = 1.93 Henries

By adding a 2H inductor as shown, we can increase the efficiency (power factor) of our circuit to nearly 100% (1.0).

▶ Representing AC Voltage

AC voltage can be represented in many ways, in the same manner that 6:45 can be represented as "15 minutes 'til seven," or as "a quarter to seven." AC voltage, being sinusoidal, has peaks and troughs, with the voltage ranging between the positive maximum and an equally negative minimum.

Although it might initially make sense to refer to AC voltages in terms of their maximums, most loads do not respond to the maximum; they respond as if the peaks and valleys are rounded off to a sort of "average" value. We use this average value, called RMS, to represent the voltage in an AC power source. Peak, and peak-to-peak values, to be described shortly, are more commonly used in signal and low-voltage applications.

Consider a sine wave with a positive peak value of 10V, and a negative peak value of –10V. The peak-to-peak voltage is the distance between the two peaks, or from 10V to –10V. This difference is equal to 20V, so we can describe the waveform as 20V peak-to-peak (20V p-p). The peak voltage is the distance from zero to the positive peak, or in this case 10V peak. The RMS voltage is equal to the peak voltage × 0.707, which, in this case would yield 7.07V RMS.

If the type of voltage specified is not written, it is assumed to be RMS. Therefore, the voltage coming out of the wall, 120V RMS, is actually 170V peak or 340V p-p.

▶ Derating a Specification by Percentage

The process of derating is essential to safety and code adherence, especially in extreme environments. When we derate a specification we subtract from it—that is to say, we make it less efficient, slower, or something less than it was before we derated it. Most specifications give a performance number that holds true at 21°C, but will also include how that performance changes over temperature. This information may be given in a chart, a table, a graph, or a derating factor. A conductor may, for example, be rated at 20 amperes (at 21°C), but for every degree above that value, 0.5% of the rated current is subtracted.

Therefore, if the ambient temperature where wiring is to be installed were 70°C (near a forge in a foundry) what current could be safely passed through the wire? First, determine the percentage we should subtract from 20A:

$$0.5\% \text{ per degree} \times 49 \text{ degrees difference} = 24.5\%$$

Nearly 25% of our current carrying capability would be lost. That means our 20A conductor now can only carry 75% of that amount, or 15A.

$$100\% - 24.5\% = 75.5\%$$

$$75\% \text{ of } 20 = 0.75 \times 20 = 15\text{A}$$

▶ Exploring the Math behind Conduit Bending

When we bend conduit, the length of raw material is less than it would be if we were to render our corners square. Assume, for example, we have a 90° bend, first with square corners and second using a 6 in. radius bend. With the square corners, the amount of material required would be 2 × 6, or 12 inches.

Using conduit bends of a 6-in. radius, we have instead one quarter circle (90°) of conduit. Using a geometry formula

$$\text{Circumference} = \pi \times \text{diameter} = 3.14 \times 12 = 37.68 \text{ in.}$$

Since this bend is one quarter of the way around the circle, the length required would be one quarter of the circumference. Therefore,

$$37.68 \text{ in.}/4 = 9.42 \text{ in.}$$

So, we are saving (12 in. – 9.42 in.)/12 inches or 22% of our conduit material when using bends rather than corners. It's also better for the wire. This savings must be taken into account when measuring a piece of conduit to be bent to fit.

Appendix D
State Licensing Resource List

This list of state contractor license boards is provided to help you find more information about licensing in a particular state.

Alabama
Electrical Contractors Board
610 South McDonough St.
Montgomery, AL 36104
(334) 269-9990 Fax: (334) 263-6115
www.aecb.state.al.us

Alaska
Alaska Department of Labor and Workforce Development
Labor Standards & Safety Division
Mechanical Inspection Section
333 Willoughby State Office Bldg., 9th Floor
Juneau, AK 99801
(907) 465-2589 Fax: (907) 465-2974
http://www.labor.state.ak.us

Arizona
State Registrar of Contractors
800 West Washington St., 6th Floor
Phoenix, AZ 85007
(602) 542-1525 Fax: (602) 542-1599
www.azroc.gov

Arkansas
Board of Electrical Examiners
10421 West Markham
Little Rock, AR 72205
(501) 682-4549 Fax: (501) 682-1765
www.arkansas.gov/labor

California
Contractors State License Board
9821 Business Park Dr.
Sacramento, CA 95827
Mailing Address:
P.O. Box 26000
Sacramento, CA 95826
(916) 255-3900 Fax: (916) 366-9130
www.cslb.ca.gov

Colorado
Colorado State Electrical Board
1560 Broadway, Ste. 1350
Denver, CO 80202
(303) 894-2300 Fax: (303) 894-2310
www.dora.state.co.us/electrical

Connecticut
Department of Consumer Protection
Occupational & Professional Division
165 Capitol Ave.
State Office Building
Hartford, CT 06106
(860) 713-6135 Fax: (860) 713-7239
http://www.ct.gov/dcp

Delaware
State Board of Electrical Examiners
861 Silver Lake Blvd.
Cannon Building, Ste. 203
Dover, DE 19904
(302) 739-4522 Ext. 204 Fax: (302) 739-4522
www.dpr.delaware.gov

District of Columbia
District of Columbia Dept. of Consumer & Regulatory Affairs
Occupational & Professional Licensing
941 N. Capitol St., NE, 7th Floor
Washington, DC 20002
(202) 442-4320 Fax: (202) 442-4528

Florida
Electrical Contractors Licensing Board
Department of Business & Professional Regulation
1940 North Monroe St.
Tallahassee, FL 32399-0771
(850) 487-1395 Fax: (850) 922-2918
http://www.myflorida.com/dbpr/pro/elboard/elec_index.shtml

Georgia
Construction Industry Licensing Board
237 Coliseum Dr.
Macon, GA 31217
(478) 207-1416 Fax: (478) 207-1425
www.sos.state.ga.us/plb/construct

Hawaii
Electrician Continued Competency Course
Honolulu Community College
874 Dillingham Blvd.
Honolulu, HI 96817-4598
PH: (808) 845-2325
http://www.hawaii.gov/dcca/areas/pvl

Idaho
Electrical Bureau, Division of Building Safety
1090 E. Watertower St.
Meridian, ID 83642
(208) 334-2183 Fax: (208) 855-2165
Code & License Info: (800) 955-3044
dbs.idaho.gov

Illinois
Department of Professional & Occupational Regulation
320 W. Washington St., 3rd Floor
Springfield, IL 62786
(217) 782-0458
www.dpr.state.il.us

Indiana
Indiana Professional Licensing Agency
302 West Washington St., Room E034
Indianapolis, IN 46204-2700
(317) 232-2980
http://www.IN.gov/pla

Iowa
Division of Labor
1000 East Grand Ave.
Des Moines, IA 50319
(515) 242-5871
http://www.iowaworkforce.org/labor/index.html

Kansas
Kansas State Board of Technical Professions
900 SW Jackson St., Ste. 507
Topeka, KS 66612
(785) 296-3053 Fax: (785) 296-3054
http://www.accesskansas.org/ksbtp

Kentucky
Office of Housing, Buildings and Construction, Electrical Licensing
101 Sea Hero Rd., Ste. 100
Frankfort, KY 40601
(502) 573-0365 ext. 125 Fax: (502) 573-1057
http://www.ohbc.ky.gov

Louisiana
State Licensing Board for Contractors
P.O. Box 14419
Baton Rouge, LA 70898
(800) 256-1392 / (225) 765-2301
Fax: (225) 765-2431
www.lslbc.state.la.us

Maine
Department of Professional & Financial Regulation
Office of Licensing & Regulation
35 State House Station
Augusta, ME 04333
(207) 624-8603 Fax: (207) 624-8637
www.maineprofessionalreg.org

Maryland
State Board of Master Electricians
500 North Calvert St., Room 302
Baltimore, MD 21202
(410) 230-6163 Fax: (410) 333-6314
www.dllr.state.md.us

Massachusetts
Bureau of Electrical Safety And Licensing
239 Causeway St., Floor 5
Boston, MA 02114
(617) 727-9931 Fax: (617) 727-5012
www.mass.gov/reg/boards/el

Michigan
Department of Labor and Economic Growth
Bureau of Construction Codes & Fire Safety
Electrical Division
P.O. Box 30254
Lansing, MI 48909
(517) 241-9320 Fax: (517) 241-9308
www.michigan.gov/bccfs

Minnesota
Minnesota Dept. of Labor and Industry
Electrical Licensing and Inspection Unit
443 Lafayette Rd. N
Saint Paul, MN 55155-4342
(651) 284-5064 Fax: (651) 284-5743
www.electricity.state.mn.us

Mississippi
Mississippi State Board of Contractors
215 Woodline Dr., Ste. B
Jackson, MS 39232
(601) 354-6161 or (800) 880-6161
www.msboc.state.ms.us

Missouri
Department of Codes & Administration
414 East 12th St.
Kansas City, MO
(816) 513-1500 Fax: (816) 513-1519
www.kcmo.org

Montana
Department of Labor & Industry
301 South Park, 4th Floor
P.O. Box 200513
Helena, MT 59620
(406) 841-2367 Fax: (406) 841-2309
www.electrician.mt.gov

Nebraska
State Electrical Board
P.O. Box 95066
800 South 13th, Ste. 109
Lincoln, NE 68509
(402) 471-3550 Fax: (402) 471-4297
www.electrical.state.ne.us

Nevada
Nevada State Contractors Board
9670 Gateway Drive, Ste. 100
Reno, NV 89511
(775) 688-1141 Fax: (775) 688-1271
www.nscb.state.nv.us

New Hampshire
Electricians Licensing Board
2 Industrial Park Drive, Bldg. 2
P.O. Box 646
Concord, NH 03302
(603) 271-3748 Fax: (603) 271-2257
www.state.nh.us/electrician

New Jersey
Department of Law & Public Safety
Board of Examiners of Electrical Contractors
P.O. Box 45006
Newark, NJ 07101
(973) 504-6410 Fax: (973) 648-3355
www.state.nj.us/lps/ca/nonmed.htm#elec5

New Mexico
Regulation and Licensing Department
Construction Industries Division
2550 Cerrillos Rd.
Santa-Fe, NM 87505
(505) 222-9800
www.rld.state.nm.us/CID/index.htm

New York
State Dept. of Labor
Division of Safety & Health
License & Certificate Unit
Building 12, Room 161
Albany, NY 12240
(518) 457-2735 Fax: (518) 485-8530
www.labor.state.ny.us

North Carolina
Board of Examiners of Electrical Contactors
P.O. Box 18727
Raleigh, NC 27619
(919) 733-9042 Fax: (919) 733-6105
www.ncbeec.org

North Dakota
North Dakota State Electrical Board
P.O. Box 857
Bismarck, ND 58502
(701) 328-9522 Fax: (701) 328-9524
www.ndseb.com

Ohio
State of Ohio Department of Commerce
Division of Industry Compliance OCIEB
6606 Tussing Road, P.O. Box 4009
Reynoldsburg, OH 43068
(614) 644-3493 Fax: (614) 728-1200
www.com.state.oh.us

Oklahoma
Construction Industries Board
2401 NW 23rd St., Ste. #5
Oklahoma City, OK 73107
(405) 271-5217 Fax: (405) 271-5254
www.cib.state.ok.us

Oregon
Building Codes Division
P.O. Box 14470
Salem, OR 97309
(503) 373-1268 Fax: (503) 378-2322
www.cbs.state.or.us/external/bcd/

Pennsylvania
Construction Regulation Dept. of General Svcs.
Public Works-Headquarters Bldg.
18th & Herr St., Room 104
Harrisburg, PA 17125
(717) 787-6371 Fax: (717) 772-3399
www.dgs.state.pa.us

Rhode Island
Department of Labor & Training
Division of Professional Regulation
1511 Pontiac Ave.
Mailing address:
P.O. Box 20247
Cranston, RI 02920
(401) 462-8580 Fax: (401) 462-8528
www.dlt.state.ri.us

South Carolina
Department of LLR
Contractors Licensing Board
Synergy Office Park, Kingstree Blvd.
110 Centerview Dr., Ste. 201
Columbia, SC 29210
Mailing address:
P.O. Box 11329
Columbia, SC 29211
(803) 896-4686 Fax: (803) 896-4701
www.llr.state.sc.us/pol/contractors

South Dakota
State Electrical Commission
302 South Pierre St.
Pierre, SD 57501
(800) 233-7765 Fax: (605) 773-6213
www.state.sd.us/dol/Boards/electrical/ELEC_HOM.htm

Tennessee
Board for Licensing Contractors
500 James Robertson Pkwy., Ste. 110
Nashville, TN 37243
(615) 741-8307 / (800) 544-7693
Fax: (615) 532-2868
www.state.tn.us

Texas
Texas Department of Licensing and Regulation
P.O. Box 12157
Austin, TX 78711
(512) 463-6599 or (800) 803-9202 in TX
www.license.state.tx.us/electricians/elec.htm

Utah
Division of Professional Licensing
160 E. 300 St., 1st Floor
Salt Lake City, UT 84114
(801) 530-6628 Fax: (801) 530-6511
www.dopl.utah.gov

Vermont
Division of Fire Safety
1311 U.S. Route 302—Berlin
Ste. 600
Barre, VT 05641-2351
(802) 479-7561 Toll Free: (800) 640-2106
Fax (802) 479-7562
www.dps.state.vt.us/fire

Virginia
Department of Professional & Occupational Regulation
Virginia Board for Contractors
3600 W. Broad St., 4th Floor
Richmond, VA 23230
(804) 367-8511 Fax: (804) 367-2474
www.dpor.virginia.gov

Washington
Department of Labor & Industries
P.O. Box 44460
Olympia, WA 98504
(360) 902-5269 Fax: (360) 902-5296
www.lni.wa.gov/scs/electrical

West Virginia
State Board: State Fire Marshal
1207 Quarrier St., 2nd Floor
Charleston, WV 25301
(304) 558-2191 Fax: (304) 558-2537
www.wvfiremarshal.org

Wisconsin
Department of Commerce
Safety & Building Division
201 W. Washington Avenue
Mailing address:
P.O. Box 7082
Madison, WI 53707
(608) 261-8500 Fax: (608) 267-0592
www.commerce.state.wi.us

Wyoming
Department of Fire Prevention & Electrical Safety
Herschier Building 1W
Cheyenne, WY 82002
(307) 777-7993 Fax: (307) 777-7119
wyofire.state.wy.us

Glossary

THE FOLLOWING GLOSSARY is meant as a tool to prepare you for the electrician's licensing exam. You will not be asked any vocabulary questions on the exam, so there is no need to memorize any of these terms or definitions. However, reading through this list will familiarize you with general words and concepts, as well as terms you may encounter in the exam questions.

A

ampacity ability of a conductor to carry current; measured in amperes

ampere fundamental unit of current flow; equal to current produced by 1V through 1 ohm

appliance device like a stove, microwave oven, or tabletop mixer. Can be hardwired, which requires no receptacles, or plugged into receptacles.

arc condition in which electrical energy jumps a gap to complete a circuit; usually occurs when voltage increases significantly or conductors move too closely together

arc blast potentially life-threatening explosion of volatile vapors (e.g., oxygen) or particles (e.g., fine sawdust) that occurs as a result of an electrical arc

arc fault redirection of current to bypass a load, usually due to frayed, split, or otherwise compromised wiring. It differs from a ground fault in that current fails to leave the circuit.

arc fault circuit interrupter (**AFCI**) device that detects and shuts down a circuit when an arc fault is detected

B

bare-wire ground ground conductors that lack insulation

bend change in conductor direction

bimetallic strip two-layered metal trigger that trips a circuit breaker

bonding establishing electrical connectivity by rendering two metallic objects into one via welding or connective devices listed for this purpose

bonding bushing entity that establishes connectivity between conduit and box

branch circuit wiring between the protective device (breaker) and the outlet or load (NEC Article 100)

▶ C

capacitor electrical component that stores voltage; used in motor start circuits

center tap method of accessing a transformer's inner (secondary) windings to provide additional voltage

circuit breaker device that opens or closes a circuit through nonautomatic means, as well as opens a circuit automatically in the event of overcurrent (NEC Article 100)

coincidental load load on at the same time as another, such as an AC compressor and a blower motor; rated at 125% of the summed current

conductor material capable of passing electrical current; opposite of insulator, or dielectric

conduit body device for pulling conductors through raceways; pull box

connected load load presented to the electrical service, as opposed to calculated load, which uses a derating factor that may or may not factor in to the connected load

contact point to which electrical energy is delivered and that is intended to connect received energy to a conductor

contactor form of relay that, through use of a coil, allows lower voltages to control higher ones; also used to start and run motors

continuous load load that is on all the time

cove lighting lighting directed upward from behind an opaque panel

current flow of electrons through a conductor

▶ D

derate to reduce ampacity based on temperature or other environmental/use factors

drip loop mechanical interface that prevents water from entering an overhead service entrance

dual-element fuse fuse with two trippable links: one that breaks a circuit quickly in the event of a short; and one that offers time delay for motors and other loads that require a surge at startup or that experience short periods of excessive current draw by design

▶ E

electricity branch of science dealing with the flow of electrons and their supply and demand

electrode terminal capable of delivering or accepting electrons

electromotive force (EMF) pressure developed by electrons wanting to get to an area where there are fewer electrons; same as voltage

explosive area area subject to fumes, vapors, or particles that, when combined with electricity, creates an explosion

▶ F

feeder conductor that connects the service entrance to the panel

finish work portion of a job done last

flexible metallic conduit (FMC) conduit made using a spiral of thin metal bands; used when vibration is a factor; sometimes called Greenfield

flush mount to attach to a surface such that the rear edge of the box is on the same plane as the surface

fuse metallic strip that melts on overcurrent, breaking a circuit. Available in standard (fast-action) and time-delay ("slo-blo") configurations.

fusible link piece of wire that melts when hot, protecting a circuit and its loads; also called a *fuse*

▶ G

gauge thickness of a piece of wire; acronym AWG sometimes used as unit of measure

ground fault situation in which current exits a circuit to ground through an extraneous path

ground fault circuit interruption (GFCI) technology that detects current leaving a circuit and disconnects the circuit quickly to prevent damage to person or property

grounding electrode conductor current-carrying device that leads to a grounding rod

▶ H

hickey imprecise type of conduit bender

horsepower another unit system for power; 1 horsepower equals 746 watts

hue pertaining to color, specifically of lighting. Incandescent lighting produces warm hues (oranges and yellows), while fluorescent lighting provides colder hues (bluish whites).

▶ I

inductor device that stores current as a magnetic field; used to smooth current spikes

insulator any material that does not allow electrons to pass through it

integrated fuse fuse built into a circuit breaker to provide additional short-circuit protection.

interruption rating measure of a breaker's ability to disconnect a circuit

isolated ground receptacle a receptacle that does not share common ground to achieve noise immunity

▶ J

junction box plastic or metal enclosure that makes splices accessible; any electrical box that is not a panel

▶ K

kitchen area where food is prepared

knockout piece of metal that is removed from an enclosure to allow wires to enter and exit the enclosure

▶ L

ladder diagram guide that depicts process control flow and function; precursor to programmable logic controller (PLC) technology

light temperature reflects color "temperature" rather than conventional temperature. Oranges and yellows are said to have warm light temperatures, while blues are said to have cool light temperatures.

limiter device that limits the maximum voltage on a line, like a safety valve on an air compressor

line voltage 120V 60Hz; most commonly used to power luminaires and appliances

liquid-tight flexible metallic conduit (LFMC) like flexible metallic conduit (FMC) but capable of protecting its conductors from moisture and water

living room area of a residence where people entertain or spend leisure time; usually a residence's largest room.

load device that converts current to some other form of energy

▶ M

magnetic element form of short-circuit protection in which a magnetic field generated by excessive current causes a breaker to trip; sensitive to magnetism rather than heat

mast portion of the overhead service that extends above the roofline; must be sturdy (thick-walled conduit) to avoid problems arising from wind and service conductor tension

megger test equipment that measures resistance to earth ground

meter base box into which a meter plugs and where service conductors terminate; plugs into a box, completing a residential circuit

▶ N

neutral conductor whose voltage is used as a reference for the voltages on other conductors

new work installation of electrical service in a new building

noncoincidental load load that is not on at the same time as another

noncontinuous load load that arises from a device that draws current intermittently

▶ O

offset piece of conduit used to slightly extend raceway from a surface to enter a surface-mounted box

ohm basic unit of resistance

old work upgrade or repair of existing electrical service

optimum power level control (OPLC) state in which a computerized controller frequently assesses demand and regulates supply to meet but not grossly exceed that demand

overcurrent condition in which the amount of current in a circuit exceeds the allowable

overcurrent protection ability of a device to protect a circuit from a more subtle increase than a short circuit; traditionally offered in residential circuits

overhead service electricity that enters a structure from utility poles; opposite of underground service

▶ P

paddle fixture box with extensions that screw into two adjoining joists; fortifies ceiling-mounted devices like ceiling fans and chandeliers

panel central distribution for a residence or zone in a large building; also called breaker box or fuse box

panelboard location of breakers and branch-circuit origination; also called a fuse box, breaker box, or electrical panel

pigtail short wire that connects an appliance to existing service; can be made on-site or purchased ready to install; often used with dryers, washers, stoves, and luminaires

pole single point of access in a panel; single pole provides 120V, while double pole provides 240V

power product of voltage and current; measured in watts

power factor logic relationship between energy consumed and energy expended to do work

primary metering measuring electrical usage on the supply side of a transformer

programmable logic controller (PLC) digital system that receives status signals and makes decisions based on these signals, energizing or de-energizing equipment in response

pull box junction box that facilitates the pulling of conductors through conduit

punchout hole in a box whose small metal or plastic cover removes to facilitate conductor entrance

► R

raceway metallic, nonmetallic, or rigid or flexible tubing that carries conductors

reamer tool that removes chatter and other sharp metal after conduit is cut

receptacle contact device installed at an outlet to connect an attachment plug (NEC Article 100)

recessed mount device mounted such that its front surface is on the same plane with the surface in which it is installed

redundancy state of having backup. In a redundant system, when a primary system (main electrical service) fails, a backup system (generator) runs.

resistance opposition to current flow; usually a measure of how much energy will be converted by a load

resonant frequency frequency at which capacitance and inductance have equal and opposite effects in a circuit

rigid metal conduit (RMC) strong, thick-walled metal pipe that can withstand physical stress

rigid nonmetallic conduit (RNC) like rigid metallic conduit (RMC) but nonconductive; often used in corrosive areas but not concrete

Romex trade name for NM-B cable, used primarily in dwellings and residential applications

rung step in a ladder diagram; usually consists of one input condition and one output condition

► S

saddle bend conduit configuration that allows two perpendicular runs to cross without connecting

service area generic term for the place where service enters a building; specifically, the area around the main disconnect and service panels in a large residential or commercial installation.

service entrance point at which electricity enters a building or structure

service entrance (SE) cable conductor that links the meter base and master disconnect or panelboard; usually of low AWG because it carries all energy in an installation

service head weatherproofed apparatus at the top of a mast that routes conductors into a structure from overhead

service point point at which the utility company cedes responsibility for a service; also called the line of demarcation

short circuit condition by which the current in a circuit is impeded only by the resistance of wire and its termination

short–circuit current rating amount of current that flows in a circuit when a load is shorted with 0 ohms

short-circuit protection ability of a device to switch off a circuit during a short-circuit condition

smoke detector device that indicates the presence of fire or smoke

splice connection of one or more conductors using solder, wire nuts, and lugs

strap generic term for a device (e.g., switch) or device set (e.g., pair of receptacles) installed in a box

subpanel panel fed from another panel. Capacity is half or less that of the supplying panel.

switch device that delivers or interrupts power to a load, making or breaking a circuit

switchgear switch panel that controls 600V or more

▶ T

thermal runaway condition in which excess current feeds high temperatures, which increase current

three-watt rule three watts (or volt-amps) of lighting per square foot of living space as required by NEC 220.12

time-delay fuse fuse that allows time to pass before tripping

time switch switch that energizes or de-energizes a circuit at a specific real time (e.g., 3 PM) or at a particular interval (e.g., on for 15 minutes, off for 45 minutes)

traveler extra wire in three- and four-way switching

trough three-sided channel for conductors

Type S fuse fuse with unique threading that forces replacement of fuses of same amperages; available up to 30 amps, applies only to 120V circuits

▶ U

underground service electricity that enters a structure from below; opposite of overhead service

utility room area where laundry and other maintenance items and equipment are stored and/or used

▶ V

volt unit of pressure or force; enough electron pressure to force 1 ampere through a 1 ohm resistance; also called electromotive force

voltage see *volt*

▶ W

waterproof prevents water infiltration within the depth-time curve

watertight resistant to water infiltration unless submerged

watt unit of power; 1,000 watts is 1 kilowatt

wattage see *watt*

weatherproof impervious to natural elements (e.g., temperature, humidity, heat, sunlight)

wire nut device that creates splices in electrical installations

wireway like a trough but enclosed on all four sides

▶ Y

yoke Y-shaped switch or receptacle that is mounted to a box; device with a yoke at either end is sometimes called a strap

▶ Z

zero point data transfer process of sending bursts of digital data during the brief period when the AC signal crosses 0 volts; transmits data via an electrical service entrance rather than separate conductors

Additional Online Practice

Whether you need help building basic skills or preparing for an exam, visit the LearningExpress Practice Center! On this site, you can access additional electrician licensing exam practice materials by using the code below. This online practice will provide you with:

- **Immediate scoring**
- **Detailed answer explanations**
- **Personalized recommendations for further practice and study**

Log in to the LearningExpress Practice Center by using this URL: **www.learnatest.com/practice**

This is your Access Code: **6840**

Follow the steps online to redeem your access code. After you've used your access code to register with the site, you will be prompted to create a username and password. For easy reference, record them here:

Username: ______________________ **Password:** ________________________

With your username and password, you can log in and access your additional practice materials. If you have any questions or problems, please contact LearningExpress customer service at 1-800-295-9556 ext. 2, or e-mail us at **customerservice@learningexpressllc.com**.

NOTES